SCIENTIFIC BIOGRAPHIES
BETWEEN THE 'PROFESSIONAL' AND 'NON-PROFESSIONAL' DIMENSIONS OF HUMANISTIC EXPERIENCES

Volume I

Series Editor
Marcin Kafar

Scientific Committee of the Series
Arthur P. Bochner (Univeristy of South Florida)
Carolyn Ellis (University of South Florida)
Marcin Kafar (Univeristy of Łódź), **Jacek Piekarski** (University of Łódź)
Danuta Urbaniak-Zając (University of Łódź)
Andrzej P. Wejland (University of Łódź)

*

The 'Biographical Perspectives' book series is a meeting place for representatives of the humanities and social sciences who situate their research practices in the auto/biographical paradigm horizon in its various guises.

The subsequent volumes will cover considerations on the theoretical, methodological and empirical aspects of auto/biographical research, as well as systematic studies devoted to an issue little explored so far, namely scientific auto/biographies. The latter research area in particular shows a huge heuristic potential. Located on the fringes of classic trends in the history and philosophy of science, sociology of knowledge and anthropology, it points to the weighty, though usually ignored or strongly marginalized personal, cultural, social, moral and axiological issues underlying the construction of the given domains of knowledge.

The publishing initiative of 'Biographical Perspectives' is also conceived as a project aimed at supporting the broader idea of the 'humanization' of science, which—in the belief of its originators—will contribute to consolidating the thought-social platform focused on creating the dialogic and autoformative space.

SCIENTIFIC BIOGRAPHIES
BETWEEN THE 'PROFESSIONAL' AND 'NON-PROFESSIONAL' DIMENSIONS OF HUMANISTIC EXPERIENCES

EDITED BY **MARCIN KAFAR**

ŁÓDŹ-KRAKÓW 2014

Marcin Kafar – University of Łódź, Faculty of Educational Sciences
Department of Educational Studies, 46/48 Pomorska St., 91-408 Łódź
e-mail: marcin.kafar1@gmail.com

Published by Łódź University Press & Jagiellonian University Press

First edition, Łódź–Kraków 2014
W.06490.14.0.K

ISBN 978-83-7969-055-8 paperback Łódź University Press
ISBN 978-83-233-3696-9 paperback Jagiellonian University Press
ISBN 978-83-7969-056-5 electronic version Łódź University Press
ISBN 978-83-233-9019-0 electronic version Jagiellonian University Press

Łódź University Press
8 Lindleya St., 90-131 Łódź
www.wydawnictwo.uni.lodz.pl
e-mail: ksiegarnia@uni.lodz.pl
phone +48 (42) 665 58 63, fax +48 (42) 665 58 62

WYDAWNICTWO
UNIWERSYTETU
ŁÓDZKIEGO

Distribution outside Poland

Jagiellonian University Press
9/2 Michałowskiego St., 31-126 Kraków

phone +48 (12) 631 01 97, +48 (12) 663 23 81, fax +48 (12) 663 23 83
cell phone: +48 506 006 674, e-mail: sprzedaz@wuj.pl

Bank: PEKAO SA, IBAN PL 80 1240 4722 1111 0000 4856 3325
www.wuj.pl

JAGIELLONIAN
UNIVERSITY PRESS

TABLE OF CONTENTS

AROUND 'BIOGRAPHICAL PERSPECTIVES'

AN INTRODUCTION

by Marcin Kafar

Institutional Background of the Project

Scientific Biographies: Between the 'Professional' and 'Non-Professional' Dimensions of Humanistic Experiences is the first book in the series 'Biographical Perspectives' developed for the English-language release. We wish this series to be treated as a global forum for the exchange of ideas meant for researchers who are willing to assign themselves to the auto/biographical thought collective in its various forms. 'Biographical Perspectives' is not only a publishing venture in the strict sense, but also and perhaps above all, a socio-scientific initiative aimed at **promoting the widely understood concept of science 'humanization' with the intention of creating an auto-formative and dialogical space**.

Such was the assumption underlying the organization, in October 2009, of the first 'biographical encounter.' It was made possible by a project Centrum Upowszechniania Innowacji w Edukacji[1] (CUIwE), implemented at that time by the Department of Educational Studies—one of the units constituting the Faculty of Educational Sciences (University of Łódź). The main theme of the said biographical colloquium (held in Łódź, at the university campus) became scientific biographies seen through the prism of the experiences of young academics rising from disciplines such as pedagogy, history, cultural anthropology, sociology, theater studies,

---

[1]   Center for Dissemination of Innovation in Education.

and philosophy. Importantly, our biographical seminar resulted in a collection of texts titled *W obliczu nowych wyzwań: Dylematy młodej kadry akademickiej* (*In the Face of New Challenges: Dilemmas of the Young Faculty* (Kafar & Kulesza, 2010)).

Eighteen months later, in the same place, again appeared an interdisciplinary group of researchers. This time we discussed an important issue of the relationship between the **'professional'** and **'non-professional'** sphere of the practice of the humanities. As the thought keystone we chose multidimensional associations emerging at the border of biographical **personal experience** and **academic experience**. The tangible effect of this conference was Volume I of the series 'Perspektywy Biograficzne' published in Polish, *Biografie naukowe: Perspektywa transdyscyplinarna* (*Scientific Biographies: A Transdisciplinary Perspective*) (Kafar, 2011). We decided to treat this monograph, since it was highly appreciated in official reviews and the feedback of the academic circles, as a base to develop a publication intended for the supralocal recipients. This task was successfully finalized two years later by completing the work on the book presented here.

The year 2012 brought another significant event in the form of a conference 'Autobiography—Biography—Narration: Biographical Perspectives in Research Practice.' The preparation of the content of the conference was participated by the representatives of the University of Łódź (including Jacek Piekarski and Danuta Urbaniak-Zając (from the Department of Educational Studies) and Andrzej P. Wejland (Insitute of Ethnology and Cultural Anthropology)) as well as globally recognizable scholars, such as Carolyn Ellis and Arthur P. Bochner, both affiliated at the Department of Communication (University of South Florida)—a leading academic center in the United States, fully focused on qualitative research.[2] The participants of the conference (over 100 persons) explored two main areas, which make essential reference planes for building up theories, drafting methodological fields, conducting analyses, making interpretations, and—last but not least—composing stories taking their origin in auto/biographical motifs. The first of these areas comprised the space of research practices applied under the humanities and social sciences, and more specifically those territories within them, which a) are distinguished by the narrated or written auto/biographies of the Other; b) concern—increasingly exposed in recent times and found especially in the layer of methodological research practices—the auto/reflexivity theme, in the extreme form manifested by a researcher referring to himself/herself as the study area (today researcher became a subject,

---

2 Department of Communication is the only institution in the United States implementing a curriculum entirely subordinated to the qualitative paradigm (source: *Autoethnography* course led by Carolyn Ellis, academic year 2010-2011).

as we would say paraphrasing a passage from a well-known manifesto (cf. Ellis & Bochner, 2000)). To characterize the second of these areas, the most appropriate expression seems to be 'scientific auto/biographies.' This little explored issue (excluding the paths of popular science and those related to them) shows a huge heuristic potential. Located on the fringes of classic trends in the history of science, sociology of knowledge, anthropology and philosophy it points to the weighty, though usually ignored or strongly marginalized personal, cultural, social, moral and axiological issues underlying the construction of the given domains of knowledge.

CB

'Biographical Perspectives,' in the opinion expressed from the position of the initiator of this project, had a chance to appear and then develop into its present form thanks to several circumstances that combine bipartite synergistic values. These are, on the one hand, specific biographical experiences (both 'professional' and 'non-professional') of a certain group of researchers, and on the other hand, a particular institutional climate, favorable for manifesting these experiences in the formula provided for in academic practice. What I mean here is fairly well illustrated by the situation in which I found myself some time ago in Kazimierz upon the Vistula River, where a nationwide Transdisciplinary Qualitative Research Seminar was held. It was there that one of my newly met colleagues made an observation that the University of Łódź is, in his view, the Mecca of qualitative research.[3] For someone who, like me, has led his whole academic life at the University of Łódź such observation was intriguing. It reminded me of the old anthropological truth—translated into the reality of scientific worlds and well exposed by Robert K. Merton (1972)—stating that an "outsider" sees more than an "insider." Perhaps, I wondered, a colleague working on a daily basis at another academic center noticed something I, remaining in the eye of the 'cyclone' of qualitative way of practicing science, was not in a position to see? Eventually, I came to the conclusion that the distinguishing feature of the University of Łódź, and more precisely—several departments of this institution cooperating in the interpersonal dimension—is in fact its openness to theoretical and methodological qualitative solutions. In this case, the faculty potential assisted by properly selected organizational tools leads to the emergence of a number of initiatives that—taken together—make for external observers a clear, multidisciplinary/transdisciplinary design. For example, at the Faculty of Economics and Sociology there is a very active

---

[3] The remark was made by Sławomir Krzychała, assistant professor at the University of Lower Silesia in Wrocław. The Transdisciplinary Qualitative Research Seminar was held in June 2012.

group of young researchers (including Piotrk Chomczyński, Dominika Byczkowska, Anna Kacperczyk, Łukasz T. Marciniak, Jakub Niedbalski, Izabela Ślęzak-Niedbalska and Magdalena Wojciechowska), who, thanks to continuous enthusiasm and sustained determination enhanced by the strong support of their mentor, professor Krzysztof Konecki, were able to create in a relatively short time, a renowned bilingual magazine with an international reach, *Przegląd Socjologii Jakościowej/Qualitative Sociology Review*. The same group of people organized The First Qualitative Research Symposium (Łódź, 10-12 June 2013). PSJ/QSR is also closely related to the Faculty of Educational Sciences, whose representatives (including Danuta Urbaniak-Zając, Marcin Kafar) sit on the Board of Reviewers of the journal and provide reciprocal conceptual help in the implementation of individual projects. The qualitative dimension of the activities undertaken within the Department of Educational Sciences comprises as well the first handbook of qualitative research in the field of Polish pedagogy prepared by Jacek Piekarski and Danuta Urbaniak-Zając (cf. Piekarski & Urbaniak-Zając, 2003). Qualitative research falls also within the domain of the Faculty of Philosophy and History of the University of Łódź, and especially of the Institute of Ethnology and Cultural Anthropology. Since the early nineties of the last century (the period of implementation of recent trends in the current anthropology and ethnology into syllabi valid in Poland), this facility has been focusing on research referring only to a small extent to classic (positivist) theoretical and methodological approaches, which is clearly revealed even in the undergraduate, graduate and doctoral theses of its graduates.

○3

A kind of 'social glue' essential for scientific communities to last are defined biographical and institutional traits. They can be distinguished also in relation to the project 'Biographical Perspectives.' Among such traits I would enumerate intergenerational and interdisciplinary continuity, which in this context seems extraordinary as it is lined with interdepartmental 'excursions' of the researchers cultivating long-term relationships. Exemplifications of mechanisms of similar provenance emerge from the biographical background of Andrzej P. Wejland and Jacek Piekarski, in connection with my own scientific biography. Fate had it that when in 2008 I completed my Ph.D. work at the Faculty of Philosophy and History I was offered a job at the Faculty of Educational Sciences. The institution that wanted to hire a cultural anthropologist (my Ph.D. thesis was rooted in that field), was the Department of Educational Studies—a unit managed by professor Jacek Piekarski, a widely known in Poland methodologist and theorist of social pedagogy. In my scientific portfolio, in addition to the standard information presenting my person, I included a certain text, which, coincidentally, was devoted to the topic of scientific biography. The article *O przełomie autoetnograficznym w humanistyce* (*On Autoethnograhic Shift in the Humanities*) (Kafar, 2010) arouse such vivid interest in my future boss that he stated during the job interview that, to use the direct quote, "he actually could give footnotes to it." As it turned out, Jacek Piekarski, which I previously had not known about, breaking the patterns of traditional thinking about the research process, wrote a piece close also to my intuitions entitled (*sic!*) *O drugoplanowych warunkach poprawności badawczej w pedagogice: Perspektywa biografii* (*On the Background Conditions of Research Correctness in Pedagogy: Biography Perspective*) (Piekarski, 2006). What is more, it also appeared that professor Piekarski many years before had been a student of my intellectual mentor, professor Andrzej P. Wejland. Both formerly explored the ins and outs of 'hard' sociology, to then go, respectively, toward the critical mainstream of anthropology of knowledge (Piekarski) and anthropology marked by 'humanistic' traits (anthropology putting man in the spotlight—Wejland). The standpoint of 'Perspektywy Biograficzne' gave us all a chance of fruitful cooperation within the same field, in the space 'between' (separate departments, disciplines, currents of thought, etc.).

Auto/Biographical Sources of the Project

Trying to determine the initial point of my biographical interests, I cast my mind back to mid-1990s. At that time, earning my first stripes in Ethnology and Anthropology (preparing to obtain a Master's degree), I came across (somewhat accidentally) Victor Turner's theory of liminality. This concept was then almost entirely unknown in Poland. It had been mentioned by a few anthropological theoreticians (cf. Burszta & Buchowski, 1992), selectively used by folklorists and ethnologists receptive to innovative interpretive solutions (cf. Wasilewski, 1989; Sulima, 1995), and only sporadically recognized by sociologists (cf. Czyżewski, 1997). The focus was predominantly put on the content of the works considered classic: *The Ritual Process: Structure and Anti-Structure* (1969) and *Dramas, Fields, and Metaphors: Symbolic Action in Human Society* (1978). In my case, the dialogue with Turner took on a slightly different form. Indeed, the erudite and conceptually catchy discoveries of the Scottish thinker were very useful for scientific writing about marginalization of people suffering from AIDS, about Polish naïve painter Nikifor Krynicki as a cultural outsider, or *Techno Culture*—marked by 'threshold' attributes (I dedicated my Master's thesis, entitled *Tematy, których mi nie odradzano: Szkice z antropologii współczesności* (*The Topics I was not Discouraged from: Sketches from Current Anthropology* (Łódź, 1997)), to the above-mentioned topics), but at the same time those discoveries were something more. Looking back, I would use phrases like the 'retrospective method,' 'treating the researcher as an object of a study,' 'observing oneself and looking at one's reflection in oneself,' 'I—the anthropologist as The Other,' etc., to describe what I attempted to do. In short, as a maturing anthropologist, I undertook an arduous and precarious task of introducing into academic discourse a strongly subjective content, revealed at different levels of experience (both *pre-* and *intra*textual). The aim of this effort was to unveil the previously hidden 'anthropological'[4] dimension of anthropology; in other words, I desired to restore a human face to the science concerning man. In *my* anthropology, I strived to deal with mechanisms underlying the functioning of social worlds, but I wanted even more to use this science to reach people of 'flesh and blood.'

Anthropology with a human face (I was strongly encouraged by my supervisor—professor Andrzej P. Wejland to practice this kind of reflection) seemed to me equally alluring and tricky. Without conceptual models at hand, I was forced to act to a large extent intuitively. I was guided by forebodings and seized every opportunity to confirm their accuracy. I experienced one of my enlightenments while reading the *Foreword* to the Aldine Edition of *The Ritual Process*. Prepared by Roger D. Abrahams it pointed out, *inter alia*, Turner's acting skills. Abrahams (1995, p. v) calls the creator of the theory of liminality a "star performer," who during his lectures impersonated with equal ease a sage, a master of ceremonies or a clown: "Espousing ideas that made newcomers expect a charismatic presentation, [Turner] insistently played the joker or the clown when he felt that his self-presentation was being taken too seriously. An academic showman, then—but one who had the ability to draw strong friendship from equally complicated people without insisting on being treated as a prophet or a star. He was so entertained, himself, by ideas arising from the actualities of group experiences, that he would rather play as master of the revels than as guru." The same author seeks the origin—and this is the key issue here—of the unique personality of the founder of symbolic anthropology in the creative atmosphere of Turner's family home. This illuminating observation[5] triggered in my mind

---

4 The words 'anthropological' and 'humanistic' in the discourse created by me gain identical tone; both—originating from Greek (*anthropos*) and Latin (*homo*)—point to the living man who cast into the world is forced to find himself in the midst of things and other humans.

5 Interestingly, this observation is confirmed by the opinion of Turner himself, who, in the *Introduction* to *From Ritual to Theatre*, published posthumously, distinctly points to close connections between his scientific discoveries and his childhood experience. These connections are well-illustrated by the autobiographical passages devoted to,

a chain of associations worth noticing here. I realized that Turner fascinated by theatre, Turner taking the role of a self-appointed actor, Turner dealing with exegesis of rituals, Turner involved in social dramas, and Turner describing those dramas in a scientific manner make up the same, inwardly *indivisible* figure. The experienced epiphany of 'life as a unity'[6] gained considerable importance. It allowed me to extract from Turner's thought the tone that greatly explains the choice of peculiar research paths followed by Victor Witter, but it also became a valuable impulse to address the problem of the place of **a personal 'I' in science**; the 'I' that manifests itself and is possible to be grasped in the variety of forms of biographism and autobiographism.

Scientific Auto/Biographies as an Emerging Field of Interest

"Why are we talking? Is it because a thinking being has something to say? But why would it utter that? Why is it not enough for it to think what it thinks? Does it not say what it thinks just because it goes beyond what is enough for this being and because language carries this profound movement?" reasonably asks Emmauel Lévinas (2008, p. 252). Undoubtedly, the discourses begin to germinate and proliferate in times of our gaining the awareness about the 'lack' of something. Auto/biographical discourse

---

e.g. the birth of the anthropological imagination (reportedly awakened by a magnificent view of the sea and sharpened by the smell of pine trees in New Forest) or to the source of specific approach to social dramas (set in emotional substratum—handed down to Victor by his mother (acknowledged actress and a founder of the National Theatre of Scotland), and analytical substratum—instilled by his father (an engineer by profession). In *The Human Seriousness of Play* Turner (1982, p. 9) writes, "My training for the fieldwork roused the scientist in me—the paternal heritage. My field experience revitalized the maternal gift of theatre. I compromised by inventing a unit of description and analysis which I called 'social drama.' [...] For the scientist in me, such social dramas revealed the 'taxonomic' relations among actors (their kinship ties, structural positions, social class, political status, and so forth), and their contemporary bonds and oppositions of interest and friendship, their personal network ties, and informal relationships. For the artist in me, the drama revealed individual character, personal style, rhetorical skill, moral and aesthetic differences, and choices proffered and made." Further on he says, "Perhaps if I had not had early exposure to theatre—my first clear memory of performance was Sir Frank Benson's version of *The Tempest* when I was five years old—I would not have been alerted to the 'theatrical' potential of social life, especially in such coherent communities as African villages. But no one could fail to note the analogy, indeed the homology, between those sequences of supposedly 'spontaneous' events which made fully evident the tension existing in those villages, and the characteristic 'processual form' of Western drama, from Aristotle onwards, or Western epic and saga, albeit on limited or miniature scale" (ibid.).

[6] To paraphrase the wording used by Andrzej P. Wejland (cf. Kafar, 2011, p. 202).

in the proposed scenario is then—*per analogiam* to other emerging discourses—an effect of feeling the lack of something. But the lack of what?

Tracking auto/biographical themes in Turner and commentators of his works (cf. Deflem, 1991; E. Turner, 1985) gradually affirmed me in the conviction that the correlations between *the experience of an individual* and *the plane of knowledge construction* are relevant. At the same time, what I found puzzling was the persistent tendency of ignoring or belittling the role of such relationships in cases where they appeared to be all too clear. In Turner's work it is possible to extract both the text fields of open looking for the sources of his own cognitive exploration and passages in which he secretly smuggled personally important themes into the official scientistic works.[7] Presumably, therefore, the author of *Structure and Anti-Structure* should say what he thought (in the Lévinas-like sense), but for some reason it did not happen. Just like it did not happen in Michel Foucault (he did not manage to realize the project "hermeneutics of the subject," the core of which was to be the idea of caring for oneself derived from Plato),[8] Claude Lévi-Strauss (who wrote *The Sad Tropics* in a moment of abandoning the pursuit of an academic career),[9] Emile Durkheim (treating his flagship work *Suicide: A Study in Sociology* in terms of an additional autobiographical

---

[7] The singularity of Turner as a researcher cultivating a kind of a 'crypto-autobiography' has been linked to his conversion to Catholicism. Viewed through the prism of the biography, the essays such as *Religious Paradigm and Political Action: Thomas Becket at the Council of Northampton* (Turner, 1978), or *Experience and Performance: Towards a New Processual Anthropology* (V. Turner, 1985), can lead to as much surprising as radical proposals. A good example of the latter are the following conclusions drawn by Mathew Deflem (1991, p. 19): "It seems that for Turner, as a pious Catholic, communitas in his later works became more a matter of faith than fact, and that he wanted to see communitas and religion everywhere leading to the day when, as Turner's former collaborator Richard Schechner explained, 'each individual will love his/her neighbor as him/herself, and when abused, will be able to turn the other cheek.' Turner's own religious experiences even led him to search for psychological basis of communitas and religion in the structure of the human brain. Thus, there was a shift in Turner's work from anthropological analysis *sensu stricto* to philosophical belief, to an attempt to look for a new synthesis 'not mainly between two scientific viewpoints [anthropology and physiology], but between science and faith.'"

[8] Cf. Eribon (2005, p. 399).

[9] Reading *The Sad Tropics* with a biographical sense was attempted by a Polish anthropologist, Waldemar Kuligowski. He writes, among other things, "The apostasy of the work of Lévi-Strauss, its 'truth,' is distinctly marked by the shallows of the biography. They gave the ultimate tone to his literary heresy, they clothed the rebellion in the robes of an escape from the positivist models to wide horizons of the possibility to choose style and find the formula that will be most suitable for one's own intentions" (Kuligowski, 2001, pp. 39-40).

utterance)[10] and Bronisław Malinowski (rightly feeling on the opposite sides of the diaries (!) that an autobiography can be transformed into a valid method of ethnographic research),[11] to name just a few of the long list of researchers 'encrypting' themselves in their scientific output.

Thus, 'Biographical Perspectives' face a very challenging task of **arousing previously 'muted' voices**; by the interpretation of the past events we want to **explore the subcutaneous meanings of texts** and to strive to show that **science is not practiced by beings abstracted from the world, but living persona—people situated culturally and socially**. Within such horizon, we will try to implement an integrated program for extracting the 'in-depth stream' of thought, the thought that can be most adequately merged by a metaphor of *the rhizome*. This metaphor probably captures best the seed status of the idea "to be hollowed," "to be deepened," "to be tested," while allowing us escape from the tyranny of general passwords, flattening the traditional scientific discourse (which was convincingly suggested by Gabriel Marcel (1952) in his *Metaphysical Journal*; cf. also, Lévinas, 1999, p. 14).

The first attempt we make is tied around the phrase '«professional» and «non-professional» dimensions of humanistic experiences.' The wording 'professional' and 'non-professional' is taken from the theory and methodology of social pedagogy coined by Jacek Piekarski (2009). As the editor of the volume, I have decided to make them central to our discussion of concepts, since they reflect well the weighty aspect of *the tension* appearing at the junction of the private sphere of our life and the scientific one. These concepts provoke us to pose questions about the relationships between scientific knowledge and the worldview of a specific researcher, the system of values professed by him/her, the field of individual experiences or personal predispositions. Personal experience of each of us ('non-professional' part of our biography) continuously—consciously or unconsciously—gets imprinted in everything that is located in the area of 'professional' actions; from the choice of research areas to the ways of their theoretical and methodological conceptualization. The process of drawing from the 'professional' experience (scientific, research) affects the 'extra-professional' life, and so we often deal with a sort of inverse feedback situation transforming me-researcher as a human being with all the resulting consequences.

---

[10] Speaking from the position of the sociologist of knowledge, Łukasz Dominiak (2008) presents an excellent analysis of *Suicide* from the perspective of Durkheim's auto/biography.

[11] The theme of Malinowski creating the foundations of autobiography as ethnography can be found in the monumental work of Michael W. Young *Malinowski: Odyssey of an Anthropologist, 1884-1920* (2004; cf. particularly *Melbourne maladies*). To read more on the analysis of the Trobriand diaries in terms of a dialogue between the public 'I' with the private 'I' cf. Kafar (2010).

From the pages of the presented book we learn how complex, difficult to predict, and simply interesting can be the fate of thoughts, entangled activities and people 'involved' in them.

∞

The way to read a work is influenced by various factors. Our texts prepared for their publication in English have undergone a gradual metamorphosis, slowly gaining the color they did not have in the original Polish version. The translation process always poses a considerable risk; translation may well 'uplift' a text or 'choke' it. The texts that make up *Scientific Biographies*, in my opinion, belong to the first of these categories. In spite of the changes that have been introduced into them, they still remain faithful to the authors who see them as their own words, despite the fact that these are the words uttered in a language 'foreign' to them.

An English-speaking reader, sooner or later, will probably realize that the work has been translated. This can be seen for instance in the bibliography, in which there have been preserved titles of works in the original language editions (mainly Polish, but also, in some cases, French, Russian and German). The same literature items appearing in the main parts of each chapter have been translated where it seemed necessary. This has been done on purpose in order to maintain the transparency of the argument. At the same time—assuming the specific competencies of potential readers of the book—we have tried to stick to the consistent use of English-language sources of citations and regular bibliographic references. The whole book has been standardized also in terms of inverted commas; we have used two types of quotation marks, single one for words and phrases that do not come directly from other authors, double one—in all other cases.

∞

This monograph would probably have not come into being without the support I have received from many people. Firstly, it seems appropriate to thank the Authors who yielded to my persuasions and spent their precious time working on their input into this publication even in moments of my, sometimes maybe a bit exaggerated, editorial scrupulosity. Special thanks go to Michał Rydlewski for his friendly support and many hours of discussions about 'science' and 'life.' My deep gratitude dating back for more than a decade goes to Professor Andrzej P. Wejland who has continued to approach my scientific-biographical ideas with patience and academic professionalism. I would also like to thank Professor Jacek Piekarski for his thorough understanding of my autobiographical texts and strong institutional support I enjoy working at the Department of Educational Studies of the University of Łódź. My gratitude in this matter also goes to Professor Danuta Urbaniak-Zając, the present Dean of the Faculty of Educational Sciences, for her trust in me during the phase of transforming this project into a supralocal undertaking. I would also like to thank Professor Grzegorz Michalski, the former Dean of the Faculty of Educational Sciences, whose supportive attitude has made it possible for the initial effects of the research on biographical perspectives to be successfully translated into the reality of the text. I thank Professors Carolyn Ellis and Arthur P. Bochner for profound talks and gestures expanding my auto/biographical self-awareness as well as intellectual and organizational commitment to our project. The momentum taken on by the English edition of *Scientific Biographies* has also been greatly facilitated by Tomasz Włodarczyk (the Director of the University of Łódź Press whose visionary idea enabled this monograph to exist abroad), Magdalena Machcińska-Szczepaniak (translating the Polish texts with great language expertise and poetic sensitivity), and also Agnieszka Oklińska (taking care, with brilliant artistic touch, of the graphic design of the publication); these persons deserve special praise for their work, whose quality goes far beyond our formal commitment. Finally, special thanks go to Marta Kozłowska for her intellectual contribution in refining my scientific thoughts and for her 'forbearing understanding' that she has granted me as the one most intensely experiencing on a daily basis what it means to deal with someone who constantly travels between 'professional' and 'non-professional' dimension of the humanities.

Włodzimierzów-Polanka / Łódź
October 2013

References

Abrahams, R.D. (1995). Foreword to the Aldine Paperback Edition. In V. Turner, *The Ritual Process: Structure and Anti-Structure* (pp. v-xii). New York: Aldine de Gruyter.

Buchowski, M., & Burszta, W.J. (1992). *O założeniach interpretacji antropologicznej*. Warszawa: Wydawnictwo Naukowe PWN.

Czyżewski, M. (1997). W stronę dyskursu publicznego. In M. Czyżewski, S. Kowalski, & A. Piotrowski (Eds.), *Rytualny chaos: Studium dyskursu społecznego* (pp. 42-115). Kraków: Wydawnictwo Aureus.

Deflem, M. (1991). Ritual, Anti-Structure, and Religion: A Discussion of Victor Turner's Processual Symbolic Analysis. *Journal for the Scientific Study of Religion, 30* (1), 1-25.

Dominiak, Ł. (2008). Między nauką i autobiografią: *Samobójstwo* Émile'a Durkheima. *Studia Socjologiczne, 3*, 75-104.

Ellis, C., & Bochner, A.P. (2001). Autoethnography, Personal Narrative, Reflexivity: Researcher as Subject. In N. Denzin & Y. Lincoln (Eds.), *Handbook of Qualitative Research*, 2nd ed. (pp. 733-768). London – Thousand Oaks – New Delhi: Sage.

Eribon, D. (2005). *Michel Foucault: Biografia.* (J. Levin, Trans.). Warszawa: Wydawnictwo KR.

Kafar, M. (2011). *Biografie naukowe: Perspektywa transdyscyplinarna.* Łódź: Wydawnictwo Uniwersytetu Łódzkiego.

Kafar, M. (2010). Zwroty zapoznane—zwroty dokonane: Autobiografia w myśleniu etnograficznym i antropologicznym. In J. Kowalewski & W. Piasek (Eds.), *"Zwroty" badawcze w humanistyce: Konteksty poznawcze, kulturowe i społeczno-instytucjonalne* (pp. 211-236). Olsztyn: Instytut Filozofii Uniwersytetu Warmińsko-Mazurskiego w Olsztynie.

Kafar, M. (1997). *Tematy, których mi nie odradzano: Szkice z antropologii współczesności.* Master's dissertation, Department of Ethnology and Cultural Anthropology, University of Łódź.

Kafar, M., & Kulesza, M. (Eds.) (2010). *W obliczu nowych wyzwań: Dylematy młodej kadry akademickiej.* Łódź: Wydawnictwo Uniwersytetu Łódzkiego.

Kuligowski, W. (2001) *Antropologia refleksyjna: O rzeczywistości tekstu.* Poznań: Wydawnictwo Poznańskie.

Lévinas, E. (2008). *O Bogu, który nawiedza myśl.* (M. Kowalska, Trans.). Kraków: Wydawnictwo Homini.

Lévinas, E. (1999). *Czas i to, co inne.* (J. Migasiński, Trans.). Warszawa: Wydawnictwo KR.

Marcel, G. (1952). *Metaphysical Journal.* (W. Bernard, Trans.). Rockliff.

Merton, R. (1972). Insiders and Outsiders: A Chapter in the Sociology of Knowledge. *American Journal of Sociology, 78* (1), 9-47.

Piekarski, J. (2006). O drugoplanowych warunkach poprawności praktyki badawczej: Perspektywa biografii. In D. Kubinowski & M. Nowak (Eds.), *Metodologia pedagogiki zorientowanej humanistycznie* (pp. 97-126). Kraków: Oficyna Wydawnicza Impuls.

Piekarski, J., & Urbaniak-Zając, D. (Eds.) (2001). *Jakościowe orientacje w badaniach pedagogicznych: Studia i materiały.* Łódź: Wydawnictwo Uniwersytetu Łódzkiego.

Sulima, R. (1995). Józef Tkaczuk i inni, czyli o imionach widywanych na murach: Przyczynek do etnografii miasta. *Polska Sztuka Ludowa: Konteksty, 2* (229), 52-64.

Turner, E.L.B. (1985). Prologue: From the Ndembu to Broadway. In E.L.B. Turner (Ed.), *On the Edge of the Bush: Anthropology as Experience* (pp. 1-15). Tucson: University of Arizona Press.

Turner, V. (1985). Experience and Performance: Towards a New Processual Anthropology. In E.L.B. Turner (Ed.), *On the Edge of the Bush: Anthropology as Experience* (pp. 205-226). Tucson: University of Arizona Press.

Turner, V. (1982). *From Ritual to Theatre: The Human Seriousness of Play.* New York: Performing Arts Journal Publications.

Turner, V. (1978). Religious Paradigms and Political Action: Thomas Becket at the Council of Northampton. In V. Turner, *Dramas, Fields, and Metaphors: Symbolic Action in Human Society* (pp. 60-97). Ithaca – London: Cornell University Press.

Wasilewski, J.S. (1989). *Tabu a paradygmaty etnologii.* Warszawa: Uniwersytet Warszawski, Wydział Historyczny.

Young, M.W. (2004). *Malinowski: Odyssey of an Anthropologist, 1884-1920.* New Haven: Yale University Press.

Chapter One

HORIZON—CONVERSION—NARRATION

IDENTITY AND OTHERNESS IN THE SCIENTIFIC WORLD OF
THE HUMANISTS

by Andrzej Paweł Wejland

Horizon

Unexpectedly for some, but with full conviction that it will be of benefit to this manuscript, I reach at the beginning—to the deliberations of Bernard J.F. Lonergan and his work *Method in Theology* (2003). I have found there a good introduction to the topic of horizons that I am interested in. "In its literal sense the word horizon—writes Lonergan—denotes the bounding circle, the line at which earth and sky appear to meet. This line is the limit of one's field of vision. As one moves about, it recedes in front and closes in behind so that, for different standpoints, there are different horizons. Moreover, for each different standpoint and horizon, there are different divisions of the totality of visible objects. Beyond the horizon lie the objects that, at least for the moment, cannot be seen. Within the horizon lie the objects that can now be seen" (ibid., pp. 235-236). The literal meaning should, however, lead me—I am not mistaken in my premonition—to the metaphorical meaning. Therefore, I read in Lonergan further: "As our field of vision, so too the scope of our knowledge, and the range of our interests are bounded. As fields of vision vary with one's standpoint, so too the scope of one's knowledge and the range of one's interests vary with the period in which one lives, one's social background and milieu, one's education and personal development. So there has arisen a metaphorical or perhaps analogous meaning of the word, horizon. In this sense what lies beyond one's horizon is simply outside the range of one's knowledge and interests: one neither knows nor cares. But what lies within one's horizon is in some measure, great or small, an object of interest and of knowledge" (ibid., p. 236).

This approach makes it possible to state that horizon—let me clarify right away that I mean here conceptual, epistemic horizons, etc.—is what defines an image or a vision of the world within our knowledge. A determined (i.e. having its boundaries) horizon is, therefore, a way to comprehend the reality and conceptualize it, dependent on how our field of vision is determined by the place in which we are situated and from which we look—our point of view. This approach, in addition to the limitation of the horizon, assumes the possibility of changing it with the shift of *locus standi*—a point which is an embodiment of ourselves in the imagined, suggested by a metaphor, sphere of knowledge and cognitive interests. This 'shift' will be of particular interest for me in the further part of this work.

Conversion as a Change of Horizon

The metaphor of a point in which one is situated, from which one looks and observes the world in a horizon, is very flexible. If we crouch in this point, we will surely see the world from a different perspective: a photographer would call it a worm's-eye view, but undoubtedly it can also be a child's view or a junior research worker's view (thus we have a metaphor in a metaphor!). And if we jump or fly up, or stand on a comfortable, paradigmatic stool, will our view not change as well? Does the academic discourse community which can live very close to the ethnographic soil (deeply rooted in empirical research) not have a different view of the world (even a 'very anthropological discourse') than the one that feels the chill of the soil and goes to the sky of philosophizing (and methodological) abstractions (i.e. the one that loves working in an office at a desk and is not keen on fieldwork)? A point of view can be also altered by: 'up—down' movement, complemented by 'left—right,' 'forward—backward,' and even 'around'—everything depends on how our 'body' (of a single researcher or the whole community) in positioned in space. The words I am using here should be interpreted with all connotations appropriate for them in our culture, they ought to be taken in their full symbolic bloom.

What is more, there is a possibility of—and now it will become exciting—a shift that leads out from a given point, a shift of one or two steps, but also of scientific miles. When a shift concerns moving from one discourse community to another, in particular moving from one paradigm to a different one, from one horizon to another horizon, especially a radically different horizon, we speak about *conversion*. Bernard J.F. Lonergan rightly argues that conversion concerns not only replacing one horizon with another, but also a change within the horizon, although a conversion through 'replacement of horizons'—as usually more 'dramatic'—probably rivets our attention more strongly. Scientific conversion is a variant of intellectual conversion, and perhaps also of moral and religious conversions (ibid., p. 237 and subsequent ones). It should be considered, as Lonergan claims, a variant of self-transcendence, of elevation that requires breaking with "often long-ingrained habits of thought and speech" (ibid., p. 239). Conversion may involve non-destructive, non-expunging memory, therefore, departing from a paradigm does not have to (and probably even should not) mean a 'biographical suicide.' It is better to choose the memory of departing over the total memory amputation—thus remembering the reasons for emigration and its path to a 'new world,' i.e. new discourse community. Only from such a dynamic perspective, one can see how the contexts are developing— the contexts that were entangling our thoughts and our vision of the world, also through the used language—and how we have enriched ourselves and our horizon.

Lonergan persistently emphasizes that conversion is something more than simply a change of horizon: conversion is also (I am using Lonergan's term here) a change of belonging: a change from one group to another or a change within belonging to the old group i.e. "one begins to belong to it in a new way" (ibid., p. 269). In the former case, it is possible to meet other 'converted' individuals, for example, converted from structuralism to phenomenological anthropology. Thus, what is also possible in a community discourse is the exchange of 'witnesses of conversion'—voices that confirm a new membership and critically dissociate themselves (but not without memory) from the old scientific environment. The emerging associations with a religious conversion are quite justified. In *The Structure of Scientific Revolutions*, quite a methodological work, Thomas S. Kuhn (2012) wrote about paradigms placed in specific research communities and did not avoid this self-suggesting analogy. For Kuhn, those communities bear (sometimes quite distinctively) features of religious cults, therefore, switching from a paradigm 'worshiped' by one community to another which is 'worshiped' by members of another community is tantamount to a conversion, sometimes very radical and violent. It is also a conversion to another 'scientific faith,' abandoning the old community, and even—from the perspective of the abandoned community—unforgivable betrayal. Shifting to another paradigm, which Kuhn called a "conversion experience" (ibid., p. 150), means not only a 'leap of faith,' but also a life-changing turnabout. Indeed, the 'leap of faith' has to have, as emphasized by Imre Lakatos, a mystical nature, and the scientific transformation resembles the religious one also because it means 'following someone'—a new leader and his supporters (cf. Barbour, 1974). For Kuhn, "transfer of allegiance from paradigm to paradigm" was a decision based only "on faith" (Trigg, 1973, p. 104).

Presenting this matter in such a way, I assume the change of conceptual, i.e. epistemic horizons and consequent 'permanent' scientific emigration, whereas a contemporary transdisciplinary, postmodern discourse emphasizes a nomad idea: traveling from one paradigm to another, from one community to another. Maybe even wandering that does not end and is—in opposition to scientific settlement and putting down roots—a permanent condition, a new lifestyle. Who is a nomad-scientist? Such person may also have—besides 'portable scientific identity'—'portable homeland,' the thing is that such person is not devoted to any community or its research paradigm with which he/she is temporary connected (if the word 'connected' is suitable in this context at all), he/she does not have the feeling of joining or entering in the full confidence the 'inside' of the community. Migration from paradigm to paradigm, if we want to use here the word 'conversion' as well, is a migration in which a researcher can have a conceptual view of conversion at his/her disposal: he/she knows what conversion is, he/she understands the essence of it, and adapts himself/herself but without real commitment, therefore, he/she usually functions in the 'as if' mode, until he/she shifts to the next paradigm and a group of its supporters. Does he/she have, in the traveling suitcase, something that he/she really identifies with, or—permanently deprived of deeper relations—he/she shall live in a peculiar state of weightlessness?

Community of Discourse and Identity

This problem could be easily resolved with the help of the concept of multiple identities, i.e. as many identities as many groups of (also scientific) participation. It is a very convenient concept releasing us from struggling with tensions and paradoxes of the identity—I mean identity understood as some unity. I choose the second option, not because I want to show how bravely I face difficulties, but because this option ensures the ability to reveal peculiar identity cracks, even in well functioning (as it would seem) humanists. Therefore, I am not interested in chameleons changing their skin color; I take into consideration more typical cases of being faithful, for a long time, perhaps even for a lifetime, to one 'scientific color.'

Can a person who is faithful (like a neophyte after conversion) to one scientific paradigm still experience some tensions —'inside' and in relations with other members of a scientific community? Let us first consider whether he/she 'writes' one story with his/her life. Kevin D. Murray (1995), following the suggestions of Mikhail M. Bakhtin and his concept of literary chrono-topes, and other scholars referring to Bakhtin, drew attention to frequent splitting of an identity into different stories. These are 'true-life' situations: we tell (or our life 'tells') something here, and we tell (and our life 'tells') something else there. These situations are also known in science and, like other situations concerning 'split identity,' were given by Murray an um-brella term of "narrative partitioning." For example, one part of the split 'I', involved in an anthropological discourse, and so a public discourse, will ad-vocate analytical psychology of Carl Gustav Jung with great (as it seems) confidence; while the second part that manifests itself, let say, in a private, banquet discourse (even among colleagues—anthropologists) can deny this approach larger scientific value; it can also be the 'I' that, 'as an anthropolo-gist' sticks to the enlightened clear-headedness and shuns all, as it calls them, superstitions, while as an 'ordinary man' avoids black cats crossing the path, going under a ladder etc. It is easy to notice that some discourses overlap and compete against one another in such 'I', and therefore, the narratives presented in different social situations can be regarded as incompatible.

The tension is born here out of conflict of some social norms functioning in the scientific world. Some social norms certainly guard narrative cohe-sion: they make the expectation that a person who, as an ethnographer and anthropologist or another humanist, sees everything—this is just an exam-ple—through the prism of sun cults, will wear these glasses also during pri-vately celebrated Easter, or the other way round—that a person who wants to spend not only Easter but his/her whole live, including his/her 'busy everyday life' as a Christian, will remain a Christian as a researcher-scholar. Other norms, demanding separation of these zones, express the expectation that the creation of a narration devoted to one of these spheres will be ac-companied by 'silence' on the part of the narration from the second sphere. It refers especially to refraining from putting 'private' matters of denomina-tion and religious faith into the public discourse. In such case, the 'silence' of private sphere narration, demanded by norms, consists in putting this nar-ration in brackets, 'freezing' it, or even performing its total—in the scientific life—'annihilation,' and therefore, one way or another, preventing it from accessing the public sphere. The researcher 'as a scholar' should not ask himself/herself certain questions at all; as a scholar he/she should not seek answers to those questions, and, above all, as a scholar, he/she should not talk about 'those matters' among the research community. For a researcher 'as a scholar' is not—or should not be—a 'layman.' Meanwhile, however, 'lay people' ("in contrast to scholars qua scholars")—as Anna Wierzbicka writes in *What Did Jesus Mean?*, a few moments before an important personal con-fession (Wierzbicka, 2001, p. 16)—will always ask themselves some vital, existential questions, for example "the question of the relevance of Jesus' teaching to their own lives" (ibid.). A personal confession, which Wierzbicka drives at in the *Introduction* to explaining in "simple and universal human concepts" Jesus' sayings, is an act, which exceedingly assures of her striving to obtain narrative coherence in her book (and probably not just there): "[...] Finally, should the reader be interested in where I personally stand, I am a believing and 'practicing' Roman Catholic. At the same time, my perspective on the Gospels has been strongly influenced by the writings of Jewish, as well as Christian, scholars [...]" (ibid., p. 23). A reader, especially a skeptical one, who does not completely believe that it is only about a scientific semantic research, may after all want to know why she took on explaining Gospel parables (instead of leaving this task to Biblical hermeneutists) and choose as her main objective even "to launch a new type of exegesis, which can be called semantic exegesis" (ibid., p. 6).

As a narrativist who is convinced but still looking for methodologically interesting perspectives, I am inspired by the thought that the division into a public and a private sphere in a scientific discourse (although the boundaries of such a division can be shifted depending on the situation) often manifests itself not so much in the 'silence' of private narrations in public circumstances, but—if, for example, issues of private religiousness are brought up in such circumstances—in using their simplified, shortened, often even trivialized versions. Meeting other researchers and the reading academic audience, under the pressure of scientific norms, can consist in their transformation into perfunctory and reticent 'summaries,' told with a language that does not suggest religious commitment too openly and into forms from which confessional elements are almost or completely leached out. Those narratives can also be replaced—by changing the rhetorical mode or genre—by humorous or self-mocking narrations defensively distancing themselves from 'serious' inclusion of private matters into public discourse. Those situationally non-developed stories should not be mistaken for narratively 'immature' stories, which are not yet arranged and remain in a prenarrative or an antenarrative phase (cf. Wejland, 2010, pp. 172-173)—regardless of whether we can process them into 'well-formed' stories or only into 'less nicely built' narratives, as *inter alia* the so-called chaos narratives (Frank, 1995, pp. 97-114; Uehara, Farris, Morelli, & Ishisaka, 2001).

What is important here is rather the way of 'editing' them—different in various social circumstances. When a person of deep faith, or a fresh convert uses such laconic, partly evasive and in a way substitutive forms of 'edition' in a public discourse, it means (I am referring exclusively to Christians) that such person does not yield to the encouragement of dauntless witnessing of God and one's religion. Anna Wierzbicka however, yielded to this dictate

(I do not go deeper into her other reasons (if such existed)—like the will to satisfy the curiosity of the readers, as she writes), therefore, *parrhesia* is a term that can be used to refer to her direct honesty (cf. Wejland, 2003). The attitude described with that term was adopted even more clearly—because it happened in an elaborated story—by Margaret M. Poloma, however, as we will see, the story was transferred to the research community welcoming such narrations.

Identity and Narration: Problem of Alienation in Science

In this way, I am driving at a problem, which does not concern only the issue of—conventionally speaking—bringing together science and religious faith, because a narrative partitioning may appear on the verge of science and any other sphere of existence (especially if the sphere is important for the 'I'). However, this field is convenient for my deliberations, because it is expressive and full of perfect real-life examples.

What has been and still is the expectation of 'science,' including the whole area of humanities, and (so to say) modern anthropology as well, i.e. a science subjected to the Enlightenment ideals? The expectation is to separate the private sphere from the public one, and avoid anything 'confessional' in *stricte* scientific deliberations. Since this expectation is generally directed against the 'professed ideology' (or anything that 'functions as a religion'), fulfilling it should not mean the agreement to including the 'professed atheism,' manifestations of irreligious attitude, or any other worldview into the research community discourse. This consequence should perhaps go even further: it should be expected that no scientific, and therefore also no anthropological or sociological, theory of religion will be based on anything ideological or religious 'inside,' and so it will not employ any scientific doctrine transformed in something like 'faith,' for example the Mircea Eliade's phenomenology of religion—translated into a kind of confession.

It is worth to remember about the criticism that has been directed at his phenomenology of religion. One of the reasons for the criticism is that this phenomenology speaks of archaic and cosmic religion and of cosmic *homo religiosus*; it equates *sacrum* with what is primitive, and being very closely connected to Jung's philosophy and holistic philosophy, for some people, it seems to become a close neighbor of New Age. Andrzej Bronk (2003, p. 291), speaking about Eliade with warm feelings, writes, "Eliade is personally involved on the side of *sacrum*, religion, and man, and his books have something deeply religious. He reiterates that a proper understanding of the world and man's ultimate goals is possible only from the perspective of *sacrum*. Eliade's religiosity is a religiosity resembling more a Taoist sage than a Christian saint. His interest in cosmic religion (earth, air, fire and water as hierophanies) reminds Ionic philosophers of Nature. The cosmic religion has no place for a transcendent and personal God and personal contact with Him. […] Eliade's ambition, however, was not to create a new form of religion, or a syncretic religion à la Eliade. He wanted to talk about religion and religious phenomena somehow beyond all theologies and dogmas. […] Eliade sees on the one hand the uniqueness of Christianity and the fact of the Incarnation of God in the history of the world, but on the other hand he repeals it, treating Christianity as an extension of the archaic religion and identifying eschatological preaching of Christ with the archaic and Indian symbolism of the rebirth of the world. Moreover, from the position of cosmic religion, Eliade is critical towards the Judeo-Christian tradition, since it—allowing the creation of particular sciences—has contributed to […], desacralization of the world criticized by him. However, he expresses the belief that the West and Christianity will get renewed through the recovery of the cosmic and sacred dimension of reality, crossing local restrictions by means of contact with Eastern religions. Eliade considers his works as contributory to achieving this task" (italics in the original).

Narrative tensions become visible after taking into consideration the fact that Eliade was brought up in the spirit of the Romanian Orthodox Church, but the journey to India, after obtaining his MA degree, immersed him into the teachings of Sanskrit and Yoga Philosophy thanks to the stay in the community of yogis in the Himalayan ashram (ibid., p. 268). When he died in Chicago, he "was provided with holy sacraments, in the presence of his praying wife and friends. The Orthodox churches in the United States celebrated the funeral prayer for his soul. The one—as Bronk writes—who used to say that the secular world is an illusion and death is a kind of initiation, went to the sacred dimension of reality, in whose existence he had always strongly believed" (ibid., p. 292).

Eliade must have felt his otherness in science, or at least in some of its areas. Here is one more passage from a thorough, extremely interesting text of Andrzej Bronk: "There are ethnologists and religious researchers who believe that what Eliade says is simply 'bad historiography, bad ethnography, bad method, bad psychology and also confusion of all concepts' (Edmund R. Leach). They deliberately ignore his work, or reject it critically as unscientific input. Other critics say that Eliade's statements are banal—who has read two of his books, has read them all" (ibid., p. 289). However, I have found a confession that weakens the above criticism in Louis Dupré's (2003, p. 13) (who writes 'as a Christian') work. This confession—at the subsoil of unreserved respect for Eliade and opposing Hegel's theory of religious negativity—builds a story of personal scientific conversion: "[...] my conversion to the dialectical view on religion has taken place entirely under the influence of positive and phenomenological research, especially the work of Mircea Eliade, and not under the influence of Hegel" (ibid., p. 14), admits Dupré.

One of more contemporary approaches to the problem of division or non-division of the 'I' and the narrative coherence or narrative fragmentation can be found in the work of researchers who insist on revealing the 'I' in anthropological narrations. Indeed, it does not always concern religion and religiousness, sometimes however, it certainly does. It is easy, while even flipping through the works of contemporary humanists, who like to refer to themselves as 'qualitative researchers,' to find stories in which they 'expose their inner self' using very personal rhetoric. For example, they treat the experience (tinged religiously) of a severe illness that struck them or someone close to them as the most appropriate subject of the so-called autoethnography. In this context, it is worth mentioning that some researchers consider 'moving out' from the world that is indifferent or hostile to personal religious commitment as the best option. They shift to the world that is friendly towards such commitment. With that goal in mind, they decide to—after intellectual conversion, but often also following religious conversion—take the path of changing the discourse community. As strangers in one community, they look for a different one in which they will be able to work without a personal split living also privately in the same non-divided horizon. A good example is Margaret M. Poloma—a sociologist and anthropologist, once a researcher of well-known 'secular' universities, after conversion to Pentecostalism (i.e. joining the *Toronto Blessing* movement) she became a member of the Association of Christians Teaching Sociology and a lecturer at the 'religious' University of Akron.

Let us first listen to what she says about the special relationship between religion and sociology in her life before the conversion (Poloma, 2001): "To paraphrase Emile Durkheim, the faith of my childhood was dying or already dead and a satisfying new faith was yet to be born. Neither Auguste Comte nor any of the masters of sociological thought could convince me that sociology could become the religion of my adult life. Although I devoured the prophetic writings of critical sociologists like C. Wright Mills and Alvin Gouldner, I was unwilling to make critical sociology (or any other sociology) my new religion. Instead I put religious concerns aside after successfully completing my doctoral prelims in the sociology of religion and adopted the stance of an atheistic existentialist and a sociological critic."

Following the conversion, those relations changed—faith did not reject sociology but required sociology to cling to religion: "Sociology was a game of life I intended to play, but I had no illusion that it would ever be worth the price of my soul. [...] I was to continue my search for integrating my faith and sociology" (ibid.).

Is it easy to practice sociology in this way, especially when it is sociology of religion? Is it still sociology of religion, or already—and still, as Danièle Hervieu-Léger (2000, pp. 9-22) would probably say—religious sociology? After all, Poloma openly claims that she wants to dabble in Christian sociology. She is aware that it is an unwelcome anomaly in the world of academic sociology: "Doing Christian sociology is much like riding a unicycle. There are many who feel that the godless discipline of sociology has nothing to say to Christians, while there are sociologists who insist that one cannot be an openly committed Christian and do good research" (Poloma, 2002a). Let us focus on (I realize that, unfortunately, it will be easy only for sociologists with deep insight into the field or transdisciplinary-oriented researchers of related disciplines), the way in which Poloma understands sociology (with no 'Christian' attribute): "Sociology represents a particular perspective or way of viewing the world. It includes a focus on how our societies and cultures are socially constructed by the people who live in them and how, in turn, these social constructs of laws, customs, and institutions 'act back' on their creators to shape and define who they are. Sociology, thus, assumes that the social world (including our religious beliefs and institutions) is created by people whose thought and behavior is shaped by that which came before them; they then modify the social world which will shape those who come after them" (ibid.).

The lack of space for extensive explanations forces me to mention only two things to illustrate the complications in sociological narratives that Poloma's *casus* may refer to.

The first issue is the presence of the name 'Holly Spirit' with different denotations and—of course—connotations, in the language of these narrations. I am convinced that Poloma refers this name to some real being (although transcendental for human reality), i.e. the third Person of the Trinity. Dealing with the witnesses of *Toronto Blessing* movement members (Poloma 1997; 2002b), she could have probably considered that she was interested only in the 'truth' of their narrations, and not the 'truth' understood in a kind of ontologically obliging way, which would have simply meant 'silencing' her of own commitment—but without its cancellation. Studying the acts of sharing the testimonies and other 'charismatic phenomena' and inquiring—after Victor W. Turner—how they create a community or how they are in favor of creating a symbolic community, she might have agreed to 'pass over in silence' the thesis that the community is ultimately created by the Holy Spirit. However, she would have firmly protested if someone had reduced the Holy Spirit to the collective consciousness, i.e. if someone had found the name 'Holy Spirit' originally empty and, wanting to re-assign (through explication) to it the referring unction, pointed to *conscience collective* as its 'real' designatum. That was the case of Matthew P. Lawson (inspired by Durkheim), in his well-known (probably also to Poloma) article entitled *The Holy Spirit as Conscience Collective* (Lawson, 1999). We easily notice how very different are the conceptual (epistemic) horizons of those two researchers—sociologists of religion. And how different are—rooted in language—the images of the world that are assumed by them and the discourse communities they are addressing their texts to, seeking mutual understanding.

The second issue is using the word 'miracle.' Margaret M. Poloma privately believes in extraordinary and miraculous actions of God in the world, and especially in his gifts—charismas. With what commitment level does she use the word 'miracle' in her texts, in which language does she read it and interpret the witnesses of other members of the *Toronto Blessing* movement? I will not answer these questions directly. Lonergan (2003, p. 222), in his deliberations concerning horizons in historic research and methodological theses of a historian Carl Becker, placed the following remark: "Can miracles happen? If the historian has constructed his world on the view that miracles are impossible, what is he going to do about witnesses testifying to miracles as matters of fact? Obviously, either he has to go back and reconstruct his world on new lines, or else has to find these witnesses either incompetent or dishonest or self-deceived. Becker was quite right in saying that the latter is the easier course. He was quite right in saying that the number of witnesses is not the issue. The real point is that the witnesses, whether few or many, can exist in that historian's world only if they are pronounced incompetent or dishonest or at least self-deceived. [...] Hume's argument did not really prove that no miracles had ever occurred. Its real thrust was that the historian cannot deal intelligently with the past when the past is permitted to be unintelligible to him. Miracles are excluded because they are contrary to the laws of nature that in his generation are regarded as established; but if scientists come to find a place for them in experience, there will be historians to restore them to history."

A little bit further on Lonergan writes that "possibility and the occurrence of miracles are topics, not for the methodologist, but for the theologian" (ibid., p. 226). Although a theologian himself, he—as a methodologist—adapts the clause of 'silence' here imposing un-touching, excluding those issues from the range of deliberations, from 'speaking.'

Dialectics of Horizons—Towards Hermeneutics

I want to use the word 'dialectics' in one of its most primal meanings, which refers to putting together notions and assumptions in a rational way, revealing the tensions or unanimity between ideas, concepts and other views. When I speak about the 'dialectics of horizons,' I mean the 'conversation' between them, which is dependent on such a combination that contains an attempt to understand one horizon within or on the basis of the other. For Bernard J.F. Lonergan, however, the dialectics of horizons means first and foremost dealing with all their conflicts—the issues that separate them and that cannot be removed by a discovery of new facts or aspects of the examined phenomena, because those issues have a rather fundamental nature: they derive from—different for each of them—explicit or implicit vision of the world, from the adopted theory of cognition, and also from the specified religious or ethical worldview, etc. (ibid., p. 236). In Lonergan's work, we immediately find a hint that this fundamental conflict exists between the two horizons in a biography (for instance a scientific biography) which is split by a conversion: namely, between the horizon which was adopted before conversion (let us name it H1) and the horizon adopted after conversion (H2). As we remember, only in particular cases conversion has a religious basis or is simply a religious conversion. In general, for Lonergan, for me, and for many other researchers, conversion means every, especially radical, change of horizon. For example, in *A Dictionary of Sociology* under the entry *Conversionism*, the author who did not provide his/her name writes, "The term can also be used in a more general sense to mean the acquisition of a new role or ideology. This general sense would embrace, for example, the idea of conversion to socialism" (w.a., 2009, p. 132).

What is now important for me is the observation that the dialectics of H1 and H2 makes the converted person go back to H1 (with which he/she is not connected by commitment anymore) from the position of H2 (which he/she is committed to due to conversion)—go back, that is include it in a biographical reflection, analyze it from the newly obtained perspective of a new paradigm. Paul Ricoeur (1988, p. 246) would call it a 'reading of oneself'—that takes into account the time and the autobiographical narration shaped according to it. During this 'reading,' i.e. interpreting, we think about "oneself as another"—the same and different at the same time, wrapped in the lines of telling one's own life, 'halved' by conversion. This 'reading' is a hermeneutic procedure: H1 and H2 belong to one person, but it is not—because of conversion—'the same' person (cf. Wolicka, 2003).

The prototype of this hermeneutic situation is, of course, 'reading of others.' H1 and H2 are properties of different people: an author of a text and its interpreter. The interpreter 'reads' the text of the author, contributing his/her own horizon to this process and trying to 'meet' the author's horizon in order to understand the text; because the condition of understanding is always pre-understanding (determined by our biographical, temporal horizon)—all that we have approaching the text we want to understand and all that we bring into the understanding of it. One cannot exclude his/her own horizon from understanding someone else and his/her text, therefore, pre-understanding cannot be excluded or 'skipped.'

In which hermeneutical situations is it easier and in which is it harder to reach such 'meeting' of horizons? I leave this question unanswered at this general level, albeit I will attempt to make some specific comments.

The proximity or distance of horizons (understood ahistorically) is the first of two issues that are perhaps the most important to me now, and I see it on a specific example. How the text is 'read' and interpreted, and how it can finally be understood and is understood by a hermeneutist, whose horizon is in tension with the horizon of the text since, for example, the text is religious in nature and the hermeneutist based his/her horizon of understanding on totally secular visions of the world and man? How, in turn, is the same text 'read' and interpreted by a hermeneutist who is a *homo religiosus* himself/herself—either from 'the beginning' or as a result of religious conversion?

There is a well-known thesis that only a *homo religiosus* can really understand experiences reported by another *homo religiosus*, i.e. the text of his/her 'spiritual autobiography,' especially if they both believe in the same thing and in the same way (to put it simply: they are followers of the same religion). For example, Simon Blackburn (2005, p. 84) elaborates this thesis in the following way: "Christian doctrine can only become evident after belief in it. The idea, especially frustrating to atheists, has echoes in the doctrine associated with the later work of *Wittgenstein, according to which immersion in a way of life is necessary for understanding its specific structures and guiding concepts."

A similar thesis is the statement that only someone experienced conversion (intellectual, moral or religious) can understand what other convert says or writes, even if it is a person 'converted to socialism,' or to hermeneutics, psychoanalysis, or symbolic anthropology in one of its variants. The supporters of this thesis insist that the experience of self-transcendence, which appears during conversion (and also encompassing oneself as another) is basically unique and non-transferable.

A certain practical aspect of this issue is also important: treating interpretations of the utterances collected by the researcher in the field as a kind of translation. Deliberating on this issue, I like to refer to (therefore, this is not the first time I have done it) Przemysław Owczarek's (2006) research in the Podhale devoted to the social and cultural image of John Paul II during the Pope's life. In this research, Owczarek did not exaggerate in his focus on the notion of image, instead, he acknowledged that the two 'pillars' of this notion are "life" and "presence." He puts these two words in inverted commas on purpose, because he means mythical life although set upon historical background, and also mythical presence similarly connected to history. The concept also contains—among other ideas—the idea of the cult of John Paul II specified according to the anthropological understanding of ritual and religion, with embedded intuitive presumption (destined for empirical, field testing) that this cult goes beyond the catholic *orthopraxis*, and even contradicts it: doctrinally erroneous and incorrect, mixed with quasi-cult behaviors typical for contemporary idiolatry, in the melting pot of 'folk piety.' Thanks to the field research, Owczarek was able to fill this whole concept with rich and vivid content, which eventually led to confirming the initial hypotheses, although they demand further specifications and development. His concept clearly engaged a specific theoretical and paradigmatic perspective, for example, he used the symbolic anthropology of Victor W. Turner to analyze the phenomena of ritual and cult placing (following Geertz's style) a dense description of a procession in the Sanctuary of Our Lady of Fatima in Krzeptówki into the frames of symbolic anthropology. As we can see, Owczarek himself did not have the ambition to create his own theory—he skillfully adapted theories of other researchers (not only Turner). How, did he, however, as an anthropologist seriously equipped with language and scientific notions and brave enough to confront a cult spreading among people living in Podhale with his own perception of this cult's doctrinal correctness, translate their utterance into this language and notions?

Generally speaking, translation is "an operation involving formulating in a language an equivalent to the utterance previously formulated in another language" (Kostkiewiczowa, 2000, p. 446). "Is it about—I asked this question before—just 'keeping the meaning'? If an interpretation is a translation, it is a special translation: in a language external to the statement it tries to reveal, but also to explain the meaning contained in it—the starting-point of interpretation is the assumption that this meaning is hidden and cannot be derived directly, but requires relevant rules of 'reading.' The result of interpretation should therefore be a text displaying the meaning 'read out' of a given utterance. Interpreting translation—let us stick to this term—not only 'shifts' a statement into a different language (which is suggested by the Latin word *translatio*), but also 'explicates' that statement in the other language (which is why we also call this procedure explication). Interpreting translation is not, as you can see, an ordinary translation: it is not about 'keeping the meaning,' but 'transcoding' an utterance into a language which—reaching into its deep levels of meaning—at the same time will explain in its own way (i.e. light up, make closer) something of what was revealed on the surface. In its own way—i.e. in the horizon proper for it, giving a new context to the sense read out from the depth, such as a context of anthropological, historiographical, sociological, and literary ideas, concepts and theories, etc." (Wejland, 2010, p. 179).

The above case, as well as other cases of interpreting translation, is related to the problem of translatability. Translatability is generally defined as "susceptibility of a text contained in a certain language to having an adequate equivalent formulated in other languages" fortunately with the added remark that "absolute translatability is only a theoretical ideal, which in the reality is replaced by different degrees of approximations only" (Kostkiewiczowa, 2000, pp. 447-448). It also concerns all interpreting translations—of religious or any other texts—conducted by a humanist researcher, including an ethnographer-anthropologist. Without excessive methodological maximalism, we can therefore ask here: whether, and to what extent such translation considers or skips, preserves or loses in the 'target' text (in our case—the anthropological interpretation) the whole deep layer of metaphors, images and symbols contained in the 'source' text (in our case—the utterances provided by informants). Is it not so that some source texts, for example texts of religious character, cannot be, as a matter of fact, translated into the language of anthropology 'without some loss'? Does anthropology not make any reductions in its 'interpretation of the interpretation'—wasting the metaphors, images and symbols of the source texts?

Perhaps, however, this problem should be put differently: is it not so that translatability is guaranteed only when interpreting translation is made by anthropologists who are 'religious people' themselves? Such formulation may be confirmed by the thesis that the language of metaphors, images and symbols—and religious language is reportedly 'by nature' such language (with the language of mysticism certainly being such language)—can only be understood through committed reading, i.e. by personal religious experience. What, however, should anthropologists who have different attitudes do? Would they be entirely or partially deprived of the successful hermeneutic access to the texts of religious character, because of their 'deafness and blindness' to deep religious meanings? This problem obviously embraces all similar cases of incongruence i.e. lack of harmony between a horizon and a language of a researcher—interpreter and a horizon and a language of an informant—author of a certain source text.

Another issue that I want to mention here is connected with taking into account, by the interpreter, the interpretation given by the author—its acceptance and 'absorption' or disacceptance and 'rejection.' In both cases, we are dealing with—as Thomas J. Csordas (1994, p. xi) says—"an interpretation of an interpretation," and we use "double hermeneutic" i.e. "a hermeneutic of a hermeneutic": we put our own understanding on the author's understanding, we, in a way, 'let the author's horizon into our own horizon'—we expand our horizon by his/her interpretation, analyze it impartially from our point of view, or negate it—with less or more commitment. This is unvaryingly underpinned by the community that shares our horizon of understanding: the view of the world retained in the language and hermeneutic consciousness of this community.

An interpretation of an interpretation can be, for example, as in Csordas' (1994) or my works (Wejland, 2004), an anthropological (i.e. developed in the anthropological community of discourse) interpretation of religious interpretations included in the testimonies of faith of the Charismatic Renewal members; and a hermeneutic of a hermeneutic can be—accordingly—anthropological hermeneutics (and not, for example, theological hermeneutics) of religious hermeneutics revealed in the testimonies (theological hermeneutics of testimonies is developed by Paul Ricoeur in one of his works (cf. Ricoeur, 1980; see also, Dziewulski, 2002)). It does not exclude more complex situations involving infusions and mergers of different interpretations and types of hermeneutics—anthropological hermeneutics with theological hermeneutics or the other way round. The language of theology often forces its way into anthropological interpretation, and theological interpretations are full of anthropological words, small ideas and quite serious concepts or theories.

This issue, of course, concerns not only Csordas and me, and not only anthropology. Many words from religious and theologians' language invade the sociological narration presented by Margaret M. Poloma. Her hermeneutics (just as Csordas' and mine), however, attempts to remain a hermeneutics that overbuilds itself over the "native exegesis" (Victor W. Turner's expression), moreover—it accepts 'native' reports or testimonies of charismatic experiences with trust or at least without questioning their veracity. In this way, she explains what it means that her research report uses a sociological approach relying on such acceptance: "This report uses a sociological approach with its strengths and its limitations to assess the effects of the so-called 'Toronto Blessing.' It is outside the sphere of the sociological perspective to call upon either God or the demons to explain what is happening in the renewal. Nor can sociology as a scientific discipline proclaim judgment about whether a given outcome is 'good' or 'bad.' (Often what is 'good' for one group of people may be 'bad' for another.) It strives for objectivity, and the information it gathers must be empirical (i.e., capable of being measured using the tools of social science). It is subjective only in that sociology relies on people to tell their stories through narrative or filling out questionnaires, accounts that are based on personal judgments. As a researcher, for example, I cannot 'prove' that people told me the truth when they claimed to be more in love with Jesus than ever before as a result of the renewal, but nor do I have any reason to be skeptical of such self reporting" (Poloma, 2002a).

In her report, Poloma does not want to leave a horizon of sociology as a scientific discipline, she, however, deprives sociology—it should not escape our attention—of the claim to 'objectivity.' At the same time, her report is not—because of her personal conversion—a text which from the start condemns itself to staying outside the academic discourse, but a text fighting for existing in the center of this discourse.

A Change of Horizon and Founding Tales

When it is time to set off on a trip to 'the new world' and when 'the new world' attracts and encourages to leave the old cognitive horizon and the language typical for it, or when instability and returns appear (just a gentle return of a previous way of thinking and speaking, blended into the *habitus* of the former community); there is a time of convincing—arguing for and against, by oneself and (even more interestingly) by others.

If the dialectics of horizons is their 'conversation,' sooner or later, a proposal can be put forward to change a horizon by the other side (frequently called 'opposite side')—to 'convert' and 'change to new faith,' for example to the faith in Sigmund Freud's psychoanalysis, Jacques Derrida's deconstructionism, or Charles Taylor's narrativism. Such a proposal attracts to its own horizon—it usually demonstrates its advantage over the horizon that should be left behind, however, it also reveals (which I will consider first) a certain founding tale. Such name is used by Erazm Kuźma (1999, p. 21), who refers to a Polish writer, Bolesław Leśmian as the 'father' of the term. Kuźma states "that at the beginning of every statement there is a founding tale fixing arbitrarily: it is so and so," and quotes Leśmian, who elaborated on this thought in his work entitled *Literary Sketches*: "The tale is never proven, always beyond the boundary of logic, gilding as if side-effectedly on the margins of experimental life—but it plays a serious role in our thinking: the role of the rainbow bridge that connects us to the illogical field of existence, the banks of the mystery, whose face is not similar to a human face" (ibid.). Following this quotation, Kuźma writes, "The theses: the era of the eye is followed by the era of language, fallocentrism is followed by histerocentrism, language creates a thing and not the other way round—these are some founding tales. Or in the rhetoric of Kuhn: this is a proposal for a paradigm shift. Or in the language of Rorty: this is a new metaphor coined to convey a new dictionary, better, or at least differently shaping the world" (ibid.).

"This founding tale, this new metaphor, the act of faith, this new paradigm" (ibid., p. 22) is placed, as Kuźma wants it to be placed, at the beginning of an argument, but it is rarely a beginning understood as *incipit*, i.e. the first words of a text. It is rather a matter of placing a founding tale at the root or in the background of an argument; often only in a mutely assumed, unutterable horizon of this argument. A founding tale is sometimes told, however only on special occasions. Does a proposal of a horizon change constitute such an occasion? Or maybe such an occasion is constituted by a situation in which a person switches his/her own horizon to ours, i.e. he/she converts and this conversion is accompanied by a rite of passage (present also in research communities)? As every rite, the rite of passage in a research community can require a reference to a myth of the origin of the community; therefore, a founding tale can be evoked in the rite as a necessary component of a collective anamnesis i.e. recalling and remembering what happened *in illo tempore*, when—let us say it—a progenitor of a research community, a 'classic' for today's members, founded a new scientific school with his/her concepts and theories. A rite of passage does not have to confirm the conversion i.e. 'initiation' into a new 'scientific faith,' because it may be only a recognition of mandatory 'coming of age' in this faith, as it happens in the case of exams, especially those which are connected with obtaining scientific or professional degrees and examine the knowledge of the fundamentals of this 'faith.' For instance, during such an exam concerning the bases of an 'ethnographic faith,' questions about the 'founding fathers' may be asked and about their views. Some questions can be accompanied by a 'silent expectation' that the 'founding fathers' will be referred to, even during the conversation about the most contemporary views on given topics. Therefore, a founding tale of Polish ethnography and its founding fathers (including Kazimierz Moszyński, Jan Stanisław Bystroń and Józef Obrębski) should always be mentioned when for example a question about an ethnographic group appears (a more contemporary ethnographer dealing with Lemkos—Roman Reinfuss should also be mentioned, but it is not enough to remember only about him).

For Erazm Kuźma, a founding tale is identical to establishing a 'new metaphor.' However, it is not just any metaphor. This metaphor is a fundamental and, at the same time, arbitrary statement: this is so, transformed into: this is it, so—according to the concept of George Lakoff and Mark Johnson (2003)—reaching a form of 'X is Y.' Such a statement has to 'assume' the whole way of thinking, seeing the world and speaking about it, so it has to be a base, root metaphor—a formative metaphor i.e. a metaphor that lays foundations for the research paradigm. As Nina D. Arutjunova (1990, p. 14) accurately notices, "The key (base) metaphors which formerly attracted mainly the attention of ethnographers and culturologists, examining the images of the world specific for different nations, in the last few decades have been found interesting by specialists in psychology of thinking and methodology of science. A significant contributions to the expansion of these issues have made by the works of M. Johnson and G. Lakoff."

Not only statements 'this is so' and metaphors 'X is Y,' but also counter-statements and counter-metaphors: 'this is not so' and 'X is not Y' are crucial from a methodological point of view—as components of founding tales. At the level of critical comments and refutation (i.e. rejection and invalidation), some argue that the old metaphor is inadequate, that conversion to the new metaphor requires complete abandonment of the old one, that we are still—it is a wording characteristic for Richard Rorty—'worshipping its corpse' when it is time to overcome the "laziness of spirit" and liberate ourselves—as Erazm Kuźma (1999, p. 25) adds—from the perseverance in the "stale language" and repeating old, uncreative metaphors as "incantations," or from using them—to refer to Tadeusz Różewicz—in constant recycling that grinds emptiness and cliché (ibid., p. 28).

Usually, encouragement to change horizons i.e. to conversion does not end with the presentation of the founding tale. Those who argue that it is worth converting often use 'true-life images.' A good image is expected to reveal, present or animate some general idea by bringing the reality closer, to 'move with the image,' in this case—an image of one's own conversion to a new 'scientific faith.' It concerns especially story-like images. In such situation, it is said that a 'colorful image' clears up the essence of a scientific turning point, makes it possible to understand the transformation, introduces into its crux by presenting the truth of the scientific turning point, especially of the accompanying doubts and mental crossroads, but also sudden inspirations and uplifting hope. Some of such narrative images assure that they contain a description of a 'representative event,' or the whole series of such events—episodes and sequences characteristic of a given scientific community as a whole (I think that Margaret M. Poloma treats her *casus* in such a way—as something characteristic or typical for many Christian sociologists). A story of those events can be a "representative anecdote." According to Kenneth Burke, "Dramatism suggests a procedure to be followed in the development of a given calculus, or terminology. It involves the search for a 'representative anecdote,' to be used as a form in conformity with which the vocabulary is constructed" (Marx, 2004, p. 96; footnote 2). It is worth noticing that the 'construction of vocabulary' means familiarizing the reality through words, including its image into language, which results in, sometimes initial, understanding. However, we should not forget that we owe this understanding also to the narrative character of an anecdote: a 'representative anecdote' is a story, and, as a story, it has a special ability of convincing 'by using an example': by referring to our life experiences and translating (or 'projecting') the experiences of other people known from their tales as well as recognizing general truths and rules (according to which others are living or used to live) on the base of those illustrations i.e. sensing and understanding the world 'from the perspective of the others,' including the scientific perspective. Do illustrations (especially the narrative ones) used in scientific communities discourse in the humanities have to be always real i.e. authentic? Can those not be imaginary (but 'moving') examples—that stimulate mind and become embedded in the heart—like *exempla* which were used as edifying disquisitions? I am leaving this question without a simple answer, because it deserves a deeper reflection, and an answer which is more complex and balanced at the same time…

References

Arutjunova, N.D. (1990). Metafora i diskurs (vstupitel'noe slovo). In N.D. Arutjunova & M.A. Zhurinskaja (Eds.), *Teorija metafory* (pp. 5-32). Moskva: Progress.

Barbour, I.G. (1974). *Myths, Models and Paradigms: A Comparative Study in Science and Religion*. New York: Harper & Raw.

Blackburn, S. (2005). *Oxford Dictionary of Philosophy*. Oxford – New York: Oxford University Press.

Bronk, A. (2003). *Podstawy nauk o religii*. Lublin: Towarzystwo Naukowe Katolickiego Uniwersytetu Lubelskiego.

Csordas, T.J. (1994). *The Sacred Self: A Cultural Phenomenology of Charismatic Healing*. Berkeley – Los Angeles – London: University of California Press.

Dupré, L. (2003). *Inny wymiar: Filozofia religii*. (S. Lewandowska-Głuszyńska, Trans.). Kraków: Wydawnictwo Znak.

Dziewulski, G. (2002). Świadectwo chrześcijańskie. In M. Rusecki, K. Kaucha, I. S. Ledwoń, & J. Mastej (Eds.), *Leksykon teologii fundamentalnej* (pp. 1189-1191). Lublin – Kraków: Wydawnictwo WAM.

Frank, A.W. (1995). *The Wounded Storyteller: Body, Illness, and Ethics*. Chicago – London: University of Chicago Press.

Hervieu-Léger, D. (2000). *Religion as a Chain of Memory*. (S. Lee, Trans.). New Brunswick – New Jersey: Rutgers University Press.

Kostkiewiczowa, T. (2000). Przekładalność [entry]. In J. Sławiński (Ed.), *Słownik terminów literackich* (pp. 447-448). Wrocław: Zakład Narodowy im. Ossolińskich.

Kuhn, T.S. (2012). *The Structure of Scientific Revolutions*. Chicago – London: University of Chicago Press.

Kuźma, E. (1999). Język—stwórca rzeczy. In S. Wysłouch & B. Kaniewska (Eds.), *Człowiek i rzecz: O problemach reifikacji w literaturze, filozofii i sztuce*. Poznań: Instytut Filologii Polskiej.

Lakoff, G., & Johnson, M. (2003). *Metaphors We Live By*. London: University of Chicago Press.

Lawson, M.P. (1999). The Holy Spirit as Conscience Collective. *Sociology of Religion, 4* (60), 341-361.

Lonergan, B.J.F. (2003). *Method in Theology*. Toronto: Toronto University Press.

Marx, L. (2004). Pastoralizm w Ameryce. In A. Preis-Smith (Ed.), *Kultura, tekst, ideologia: Dyskursy współczesnej amerykanistyki* (pp. 95-132). (M. Paryż, Trans.). Kraków: Towarzystwo Autorów i Wydawców Prac Naukowych Universitas.

Murray, K.D. (1995). Narratology. In J.A. Smith, R. Harré, & L. Van Langenhove (Eds.), *Rethinking Psychology* (pp. 179-195). London – Thousand Oaks: Sage. Online version of the text: *Narrative Partitioning: The Ins and Outs of Identity Construction*, http://home.mira.net/~kmurray/psych/in&out.html [last accessed: June 21, 2003].

Owczarek, P. (2006). *Karol Wojtyła—Jan Paweł II: Podhalańska opowieść o świętym: Od historii do mitu—studium antropologiczne*. Kraków: Wydawnictwo Avalon.

Poloma, M.M. (2002a). *Fruits of the Father's Blessing: A Sociological Report*, http://www3.uakron.edu/sociology/FRUITS.pdf [last accessed: June 7, 2005].

Paloma, M.M. (2002b). "Toronto Blessing" [entry]. In S.M. Burgess & E.M. van der Maas (Eds.), *The New International Dictionary of Pentecostal and Charismatic Movements*. Grand Rapids – Michigan: Zondervan.

Paloma, M.M. (2001). *Pilgrims' Progress: An Exercise in Reflexive Sociology*, http://hirr.hartsem.edu/research/pentecostalism_polomaart9.html [last accessed: October 10, 2013].

Paloma, M.M. (1997). The "Toronto Blessing": Charisma, Institutionalization, and Revival. *Journal for the Scientific Study of Religion, 2* (36), 257-271.

Ricoeur, P. (1988). *Time and Narrative*, Vol. 3. (K. McLaughlin & D. Pellauer, Trans.). Chicago: University of Chicago Press.

Ricoeur, P. (1980). The Hermeneutics of Testimony. In L.S. Mudge (Ed.), *Essays on Biblical Interpretation* (pp. 119-154). Philadelphia: Fortress Press.

Trigg, R. (1973). *Reason and Commitment*. London – New York: Cambridge University Press.

Uehara, E.E., Farris, M., Morelli, P.T., & Ishisaka, A. (2001). "Eloquent Chaos" in the Oral Discourse of Killing Fields Survivors: An Exploration of Atrocity and Narrativization. *Culture, Medicine and Psychiatry, 25*, 29-61.

w.a. (2009). Conversionism [entry]. In J. Scott & G. Marshall (Eds.), *A Dictionary of Sociology*. Oxford – New York: Oxford University Press.Wejland, A.P. (2010). *Dyskurs i tożsamość: Opowieści we wspólnocie naukowej*. In I. Taranowicz & Z. Kurcz (Eds.), *Okolice socjologicznej tożsamości: Księga poświęcona pamięci Wojciecha Sitka* (pp. 155-182). Wrocław: Wydawnictwo Uniwersytetu Wrocławskiego.

Wejland, A.P. (2004). Wspólnota świadectwa: Charyzmatyczne opowieści o uzdrowieniu. In G.E. Karpińska (Ed.), *Codzienne i niecodzienne: O wspólnotowości w realiach dzisiejszej Łodzi* (pp. 29-77). Łódź: Polskie Towarzystwo Ludoznawcze.

Wejland, A.P. (2003). Parrhesia—piękne i mądre słowo. *Szum z Nieba, 4*, 26-27.

Wierzbicka, A. (2001). *What Did Jesus Mean? Explaining the Sermon on the Mount and the Parables in Simple and Universal Human Concepts*. New York: Oxford University Press.

Wolicka, E. (2003). Odkrywanie tożsamości "Ja"—hermeneutyka Paula Ricoeura "w drodze" ku fenomenologii osoby. In A. Grzegorczyk, M. Loba, & R. Koschany (Eds.), *Horyzonty interpretacji: Wokół myśli Paula Ricoeura* (pp. 99-111). Poznań: Fundacja Humaniora.

Chapter Two

BIOGRAPHICAL EPIPHANIES IN THE CONTEXT OF LAYING
THE FOUNDATIONS OF THE QUALITATIVE THOUGHT
COLLECTIVE

by Marcin Kafar

Introduction

In May 2006, the University of Illinois held the Second International Congress of Qualitative Inquiry conference. One part of the congress was a panel moderated by Carolyn Ellis, perhaps the most prominent promoter of
the autoethnographic trend in the contemporary humanities. She invited
to the meeting several leading figures involved in the qualitative research
methodology, including Norman K. Denzin, Yvonna S. Lincoln, Laurel
Richardson, and Arthur P. Bochner. These researchers, supported by
Ronald J. Pelias and Janice M. Morse, made an attempt to portray the
context of the creation of a thought collective, tied around the qualitative
thought style.[1] The palette of issues streamlining the individual statements
(presented in the form of several-minute flashback stories) mostly fluctu-
ated around the lives of individual researchers, from whom they learned
the qualitative approach, which events conditioned their location in such
and no other point of the scientific scene, and finally, where on this scene
they situated themselves.

The narrated stories (contained in one of the chapters of the book *Ethical
Futures in Qualitative Research*)[2] bring closer the political and institutional

---

[1] The concepts of 'thought collective' and 'thought style' I use in the senses given to them
    by Ludwik Fleck (1979 [1935]).

[2] *Coda: Talking and Thinking about Qualitative Research* (Ellis et al., 2007).

background of qualitative thinking patterns; the imagination is also moved by—complementary to the layers describing the outer conditions of the creation of the qualitative thought collective—these planes of the stories, in which we find direct references to the experience that is *personal*, sometimes *intimate*, to the breakthroughs and decisions following them, both 'professional' and 'private'—purely human.

The existence of these shades of biographical stories, which is clearly shown by the accounts referred to in the subsequent paragraphs of this chapter, suggests that without the individual experience (and the specific type of self-reflection built on it), it probably would not be possible to constitute the field of qualitative research in the form in which we know it today; this conclusion leads in turn to a wider question about the role of **the personal dimension in the context of creating scientific worlds**.

Years ago, Ludwik Fleck (2007a; 1986a; 1979 [1935]), followed by Thomas Kuhn (1977; 1996), clearly and convincingly exposed the social nature of the construction of scientific knowledge. At the same time, the same authors, eagerly supported by masses of commentators of their work, effectively assured us in our belief in the fact that what allows knowledge to take on an expressive and relatively permanent form (thought style, paradigm), is the close dependence on certain group activities. Knowledge gains legitimacy circling within the web of intersubjective relations; it is a *sine qua non* condition for practicing science at all. While not acting against this in a sense unquestionable principle (extreme individualism is essentially non-discoursive), it is worth while to think to what extent, within what scope, and with the mobilization of what resources an individual actually participates in the process of building collective consciousness.

This trail arises somehow involuntarily when we start to cope with the confidences of Denzin and Lincoln, and, for some reasons, first of all, with the considerations of Bochner, Ellis and Richardson. What is collective, what aspires to be a turning point in the community dimension, to constitute, as it would be put by Fleck, a "form"[3] that organizes *the* collective,

---

3   In Fleck's theory of thought styles and facts, the word 'form' is one of the most important words. It is a metaphor embracing the distinctive features of some collective: objects of observation and ideas related to them. "We look with our own eyes, we see with the eyes of a collective body," Fleck (1986b [1947], p. 134) used to say. Drawing on the findings of psychology he described form as follows: "[...] every perception is, in the first place, a seeing of some wholes, while the elements are only seen later. Sometimes these elements may remain unknown. We recognize at first sight a man of our acquaintance or a known flower, but often we are completely unable to give the distinguishing features accurately. We see all at once that somebody has a sad look, though we do not know which detail of his facial features changed. We see that the general appearance of a room has changed, but we do not know which item of furniture have been moved. Moreover,

grows from the soil of life's dramas—the death of the loved ones, accident, serious illness, accompanying others in suffering, contact with the evil deepen the old and raise the new doubts, symbolically activating in the researcher-human the desire to radically rethink the attitude towards the world in its various forms, and in different revelations: theoretical, empirical, experiential, emotional, relational, creative, and that reflected by all the others, forming me as a person and being a value in itself.[4]

It is understandable that practicing science deposited on such basis, they had to adopt separate, characteristic *for them* criteria for the delimitation of cognitive research boundaries, objectives, methods, and techniques for—to use the old language—collecting empirical data; they also dared to develop peculiar interpretation tools and methods for their use; and finally, they made the effort of systematic work on the contemporary vocabulary and rules for its use, oscillating towards writing corresponding to the needs of the times of change (the subject of language will be referred to a little more fully below).

The effect of these actions was cementing the platform, whose construction was initiated by a group that originally was quite scattered, and then more and more consolidated, as they used to talk about themselves. It was a group of scientific outsiders, individuals perceiving the reality differently than those inclined to maintain the *status quo* representing the mainstreams of the social sciences and humanities. This revolution could happen due to the favorable 'spirit of the times' and "liminal *personae*"[5] feeling it.

---

in spite of many different details we can observe an identical form within a specific whole, a specific 'entirety'; thus all Chinese people may seem to be identical to the eye of a European, although undoubtedly they have individual differences. The word 'father' when pronounced by the squeaky voice of a child and by the drunkard's bass of a sailor may have not a single sound in common, but it is still the same word. It is precisely such entireties, which trust themselves upon sensory perception, and which are to a large extent independent of constituent elements, that psychology calls 'forms', regardless of the sense which supplies them. Thus we can have visual forms, e.g. a certain tune, a word; or olfactory ones, e.g. the smell of grocers' shops, or of railway stations" (ibid., pp. 130-131)].

[4] I base this idea on the writings by a Polish philosopher Józef Tischner, discussing in his works on dialogue the notion of 'axiological I' (see e.g. Tischner, 2005; 2006a; Tischner & Kłoczowski, 2001). The associations between Tischner's ideas and the qualitative thought style perspective in autoethnographic manifestation are described in *O przełomie autoetnograficznym w humanistyce* (*On Autoethnographic Shift in the Humanities*) (Kafar, 2010), which, in its thematic content, is strongly intertwined with the text presented here.

[5] Expression borrowed from Victor Turner (1978).

Biographical Epiphanies

Story I—Norman Denzin and the 'Skipped Line'

"By the end of the 1970s, I had hit a brick wall and other walls as well. I had taken symbolic interaction about as far as I thought it could take me. And I was profoundly dissatisfied with the wall the perspective had hit. That is, it had become closed off from all sorts of other discourses that I was being exposed to on this campus in criticism and interpretative theory, which was an inter-disciplinary program in the humanities.[6] So by the late '70s, we were reading European social theory that was just being translated. Lacan, Heidegger, Foucault, the feminisms, and we were moving into semiotics. It was a three-year project of being saturated with theory that sociology was excluding.

About this time, we formed a traveling minstrel show and some of the members are on this panel: Carolyn, Laurel, myself, and Patricia Clough. We would go to the symbolic interactionist annual meetings and do post-modern performances, and we would get booed and hissed. **One of the more profound moments was when Laurel presented the life of Louisa May**,[7] the poetic representation of an interview transcript. She later published this as 'The Skipped Line' (Richardson, 1993). The room was like this, packed, and Laurel had distributed her transcript of this interview, which she then proceeded to poetically perform for us. **I think it was Harvey Farberman**, a symbolic interactionist, **who raised his hand and said, 'You skipped a line, and therefore, the validity of what you are doing is at question. You are not being true to her life and to her words.' That skipped line provoked a give and take in the journals and opened this space that we were in, a space of skipped lines, and it was okay to be there**. Even if our colleagues didn't like it, that was the space we were going to be in. So then for several years, we did this kind of traveling road show and confronted a fair amount of hostility. But as we did, I think the momentum started to build behind us" (Ellis et al., 2007, p. 240).

Story II—Yvonna Lincoln and an Encounter with Evil

"I want to talk about an epiphany and an ethical crisis at once. But before I do it, you should know that I lived a sheltered childhood. I grew

---

[6] Denzin describes a situation that occurred at the University of Illinois (footnote—M.K.).

[7] The total of the phrases in bold in this section is intended to provide a kind of an over-text, a tale within the tale that keep a dialogue with each other and complement each other to form a whole.

up in a very traditional family. My brothers ran wild. I was locked in the house for thirty-two years. So I didn't have much experience with a lot of stuff, and **my epiphany came when I was out on an evaluation contract**. I was very new. I had had my doctorate for about a year and I was doing this evaluation contract and trying to be a good qualitative naturalistic evaluator. **I began to suspect that in this project, the middle school coach was molesting some of the boys on one of his teams. And I didn't know what to do about that**. That was before we knew very much about laws that said you had to report stuff. I went to the superintendent and said, 'I don't have any hard evidence, but there is a bunch of kids who are telling me things and I think you need to do something about it.' **Once the young boys were questioned, it turned out to be true. I thought all qualitative researchers were all good and that they encountered only other good people in the world.**

My epiphany was finding out there really is evil in the world; there really are hideous things that happen to kids that never should. I have to tell you that that came as an epiphany to me because it was the first time I realized that. I know that really sounds stupid—see this is one of the things that doesn't go in, right?—but this was the first time in my life that I felt that I had come face to face with what I geminately would describe as evil rather than bad or rude or discourteous or un-Christian. It was evil, just evil. I was very young; I don't think I was but thirty-one years old. **It was quite frightening, and that was an epiphany for me**" (ibid., pp. 240-241).

Story III—Laurel Richardson, Coma and Split 'I'

"I love the word epiphany. Every part of your mouth gets going: e-pi-ph-any. Marvelous word. And I think Norm is the one who's introduced it into our living research vocabulary. I want to quickly talk about two of my epiphanies.

I've always been a qualitative/quantitative researcher—what is now call mixed methods. I find them kind of fun. Mixed-up methods. But **I had a major car accident and I was in a coma for some while and I when I came out of that coma, I was not able to do my fourth-grade mathematics. That was a life-changing epiphany**. I lived through that, but my mind was pretty scrambled up. My first paper was a power analysis of *Paradise Lost* (Milton, [1667] 2003). I hadn't read Paradise Lost since college. I didn't even know I knew it, but I did. Things were scrambled up, but I wanted to continue being an academic. My department didn't want to tenure me because I might be brain damaged (little did they know what was yet to come). At that time, feminism was growing but there was no structured way of teaching students about gender issues. I was a feminist, so I decided to write a textbook, *The Dynamics of Sex and Gender* (Richardson, 1977). Some of you may have read it. Writing that book introduced a new field and helped establish it as one that did not require knowledge of statistics to make sense of the world. That is, it established gender studies, women's studies, and sociology of women as fields of knowledge accessible to everybody. People without advanced mathematics. And through the writing, I retaught myself the bases of sociological reasoning. Writing for my life, writing so I would have a life.[8] That was the first epiphany.

The second epiphany was having a book contract on unwed mothers and finding myself unable to write. I was frozen. The crisis of representation had truly hit me. I didn't know how to write. For whom do I write? Whose life can I write? What do I say? At the same time, I was experiencing the tension between two sides of myself: the scientist and the poet. I wanted to feel more integrated. How was I going to put myself together?

I ended up writing a life-history interview of the unwed mother, 'Louisa May,' as a narrative poem and presenting it a sociology conference. **This experience**—along with talking with others who were also involved with writing themselves out of the crisis of representation—**created the space in the discipline and in our world where we could be a community**. People who were interested in altering qualitative methods, who recognized poststructural thinking, postpostmodernist critique, feminism, queer theory, and so on, could now have a space in which to create community. **The experience of performing Louisa May at an ASA convention, where people swore at me and accused me of fabricating my research, led to my involvement with others who had their epiphanies in the same space**" (Ellis et al., 2007, pp. 242-243).

Story IV—Arthur Bochner, the Death of His Father and Meeting Carolyn Ellis

"Before I read Norm Denzin's book (1989), which focused on epiphanies, I didn't know what an epiphany was. Now I see them everywhere. **In 1988, my father died suddenly of a heart attack**. I was at an academic conference, a National Communication Association convention in New Orleans. And my world was shaken by that experience. I wrote about that in a story I call 'Narrative and the Divided Self' (Bochner, 1997). **My father's death exposed to me the cleavage of my experience as a human being and as an academic. I had always struggled with this distance between the personal and the academic. I realized after my father's death that my days were numbered, and it was time to stand up and do what my heart said was important.**

---

8   More information on these issues can be found in e.g. *Getting Personal* (Richardson, 2001).

At the time, I was in somewhat enviable position of being the chair of the USF Communication Department and **I felt it my calling**, I guess, to develop a new Ph.D. program that very much embodied what Laurel just said about a sociology without quantities. I never believed that communication was the stuff of quantities to begin with. At that time, in 1990, we had the opportunity of developing this program that everyone in my discipline and even many in my home university said would never work. But I was firm in my conviction that there were people out there, especially woman, Third World people, and indigenous populations, who were yearning for such an opportunity.

I also **had a serendipitous meeting with Carolyn Ellis in 1990**. I attended a lecture she gave in, of all places, the business school, and as I heard her give a short narrative taken from her book, Final Negotiations (Ellis, 1995), I said to myself, 'She's giving my talk. There's someone else out there in another discipline who believes all the things I believe.' **That**, as they say, **was the beginning of a beautiful friendship and the start of our project on ethnographic alternatives**" (Ellis et al., 2007, pp. 243-244).

Story V—Carolyn Ellis, Life Soaked with Loss and Finding 'the Other I'[9]

"**My first epiphany occurred when my brother Rex died in 1982 in an airplane crash on his way to visit me. My world was turned upside down,** and I think this was the only time in my life I would define myself as depressed. **Not only had my brother died but my partner, Gene Weinstein, was entering the final stages of a chronic disease. The survey study on jealousy I was doing seemed insignificant and I craved to explore and try to understand what I was feeling—to get myself out of the depths of despair. That was the beginning of my turn to autoethnography, to exploring and writing about myself and my situation to learn about human behavior.** Finally, I was able to connect my love for social psychology with my love for engaged qualitative methods.

My second epiphany came with the death of my partner, Gene Weinstein and the responses I got to my writing Final Negotiations about relationship and his dying. I felt the narrative story I was writing was the best sociology I had ever done, and **to get the varied responses I received was mind-opening and mind-boggling: 'This isn't sociology or research,' 'This threatens the whole sociological enterprise,' and so on. All of it made me more determined to make my case that this was sociology.**

---

[9]   „For Art, my other I," this is the dedication put by C. Ellis in her programmatic book, *The Ethnographic I: A Methodological Novel about Autoethnography* (2004).

Norman's response to a paper I did on introspection[10]—that I was being schizophrenic—helped me move from trying to fit into a mainstream sociological model to finding my own place on the margins, one
that connected humanities and social science and advocated for an emotional sociology that cared about people.

My third epiphany occurred when I met Art Bochner and found
a like-minded colleague and partner. Together we created a synergistic
relationship and ethnographic project, and we were able to do more together to advance an interpretive and humanistic social science than
either of us could have done alone" (ibid., pp. 244-245).

On Understanding the Meaning of Epiphany

The readers of the stories presented above received a clear indication for
interpretation—that is the word 'epiphany,' and more specifically, the meaning layer of that word. The intention of Carolyn Ellis preparing the thematic structure of the panel was to treat epiphany as a 'headword,' designed
to refresh the memory of the congress participants. This solution was the
result of inspiration derived, as suggested by Richardson and Bochner, directly from the thoughts of Norman Denzin. Therefore, it is worth, I think,
bringing the latter closer, treating them as a prelude to further discussion.

The author of *Interpretive Interactionism* harnessed the concept of epiphany for biographical analyses, highlighting the conscious participation
of human subjects in the social world. We focus here on persons experiencing **moments of crisis**, interpreted—through incorporating them into
the plot of biographical and autobiographical stories of life transforming
moments—as **turning point experiences**. The flagship *exemplum* of such
turning point crisis is the life of Martin Luther King. On January 27, 1956,
the activist fighting for the rights of African-Americans received threaten-
ing phone calls. Later, not being able to sleep, he kept sitting at an empty
kitchen table and wondered about what lay ahead. At one point, lost in
the depths of despair, he heard—as he claimed—the voice of Jesus ensuring
him that he will never leave Martin: "I could hear an inner voice saying to
me, 'Martin Luther, stand up for righteousness. Stand up for justice. Stand
up for truth. And lo I will be with you, even until the end of the world.' ...
I heard the voice of Jesus saying still to fight on. He promised never to leave
me, never to leave me alone. No never alone. No never alone. He promised
never to leave me, never to leave me alone," says King. This scene is recalled, *inter alia*, in the book, *Bearing the Cross* by David Garrow (2004, p. 58),

---

[10] Cf. Ellis (1991).

which strongly stresses its **transformative nature**. The biographer of the future Nobel Prize winner notes that "the vision in the kitchen" repeatedly became a reference point for King, especially when the burden he carried seemed too heavy for him (cf. ibid., p. 89, 123, 171, 412).

Denzin willingly uses the example of Martin Luther's biography, introducing the reader into the context of his own understanding of epiphany. Perhaps this is because the *casus* of King allows him to make a relatively smooth transition between the source semantic plane, pointing to the **illumination**[11] as a specific form of establishing the relation between God and man (the chosen person), and a revelation not having a clear religious connotation. Social researcher is interested not so much in capturing the experience limited to the sphere of what is felt, generally speaking, in the purely spiritual dimension, but rather in the separation of the principle (and using it as an interpretation trick) of the **'discovery' of what had previously remained 'overshadowed.'** Perhaps this is why, in my opinion, the term 'epiphany' first used in *Interpretive Interactionism* is provided with a quotation mark (Denzin, 1989, p. 15), and as an additional support—making the intentions of the hermeneutist more precise—there emerges James Joyce with his *Dubliners* (ibid., pp. 16-17), a text which serves as a field for experimenting with the original theory of epiphany.

The touch of Joyce is really well suited within the contour of the interpretive approach developed by Norman Denzin, since, in fact—it is my additional stance—he needs 'new' clothes for the 'old' experience; the genius of the creator of *Ulysses* is proving to be invaluable in this regard, which seems to be confirmed by the expert in Joyce, i.e. Tomasz Gornat. In his erudite work, *"A Chemistry of Stars": Epiphany, Openness and Ambiguity in the Works of James Joyce* (2006), he looks for the foundations of the theory of epiphany in the utterances of Stephen Dedalus, who, as we know, is Joyce's literary disguise. In *Stephen Hero*, Dedalus defines epiphany as "a sudden spiritual manifestation, whether in the vulgarity of speech or of gesture or in a memorable chase of the mind itself" (Joyce, 1963, p. 211). Such 'spiritual raptures'—leaving Joyce aside for a moment—belong to the two spheres, corresponding to the heterogeneous nature of epiphany: first, in

---

[11] Epiphany was originally rooted in the Greek cultural context, which well reflects the language and its use (*epipháneia* means a 'revelation' from *epiphaínein*—'to show,' 'to reveal' (source: *Wielki...*, 2010, entry: epiphany)). In ancient Greece, epiphany exclusively meant literal illumination. Gods and goddesses descending to Earth and interacting with mortals were described by the then contemporary literature (such as the *Iliad*, *Odyssey* and less well-known texts, such as the *Apollo Epiphanies* by Istrus). In biblical terms, epiphany (and theophany) embraces all situations in the world of the divine revelation in the world of man (cf. Langkammer, 1990, p. 53). Also, the Christian feast celebrated on January 6 is known as the Epiphany.

the case of consciousness focused on religious feelings, man directly experiences things that bear the hallmarks of divinity (**theophany**); and second, in the case of awareness facing towards sensations deprived of the divine element, we deal with **illumination**, equipping us with the brand new 'vision' of things, flashing in the previously unknown light tone[12] (**secular epiphany**) (cf. Gornat, 2006, pp. 23-24). Joyce's Stephen reveals to his interlocutor that the clock on the Ballast Office may also cause an epiphany: "I will pass it time after time, allude to it, catch a glimpse of it. It is only an item in the catalogue of Dublin's street furniture. Then all at once I see it and I know at once what it is: epiphany" (Joyce, 1963, p. 211). Epiphanies are Virginia Woolf's "little miracles," her "matches struck unexpectedly in the dark" (Woolf, 1960, p. 249), Conrad's "moments of awakening and vision" (Conrad, 1921, pp. vii-xii), Hemingway's "moments of truth" (cf. Beja, 1971, pp. 49-52) and Eliot's "timeless moments in time" (cf. Nichols, 1987, pp. 190-198).

Attributes of Epiphanic Experiences

The distinction between the divine insights into the earthly world of people from secular illuminations is helpful both in organizing Denzin's ideas and making exegesis of the stories that I am most interested in, of scientific transformations and their sources. At least at the first glance, the stories of Lincoln, Bochner, Richardson, Ellis and Denzin, have distinguishing features allowing to locate them on the side of secular epiphanies. To give the reader a closer hint of what I mean, I will refer to (based on the definition of epiphany of Morris Beja,[13] tracking theophanies and illuminations in modern literature) three categories, or rather principles that make up a kind of **idiom of epiphanicity**, in terms of: (i) incongruity, (ii) suddenness[14] and

---

[12] I am playing with the meanings of the source word; *epiphaínein* can be translated as 'to reveal' or 'to light up.'

[13] The definition I refer to describes epiphany as "a sudden spiritual manifestation, whether from some object, scene, event, or memorable chase of the mind—the manifestation being out of proportion to the significance or strictly logical relevance of whatever produces it" (Beja, 1971, p. 18). This definition, as can easily be noticed, largely coincides with Joyce's definition put into the mouth of Stephen Dedalus; it complements the latter adding to it the rule of disproportionality. To read more on the comparison of the definitions coined by Beja and Joyce see Gornat (2006, p. 26).

[14] I'm talking about the 'suddenness' for the lack of a better word, plastic enough to fully embrace the specificities of epiphanic experience in its *initial* point. When we are on the border of the Words and Non-Words dimensions, it seems most appropriate to think—again I follow Józef Tischner—"from within the metaphor" (Tischner, 2006c, p. 240); epiphany is an 'instantaneous flash of fire,' 'first ray of light falling into the cave,' these "matches struck unexpectedly in the dark" (Woolf).

(iii) spirituality, all of which are written into an epiphany and determine its specific character. The descriptions, which I will refer to in a while, embrace each of these rules. What are the consequences that result from it?

Suddenness

"One of the more profound moments was when Laurel presented the life of Louisa May, the poetic representation of an interview transcript," Denzin-narrator says, recalling at the same time an assessing statement of Harvey Farberman ("You skipped a line, and therefore, the validity of what you are doing is at question.") and concluding: "That skipped line provoked a give and take in the journals and opened this space that we were in, a space of skipped lines, and it was okay to be there." Farberman's voice acts like a surgical scalpel, in a split of a second dividing the reality and causing mutual surprise. His casual remark reveals that there is no 'us', but there is 'we' and 'they'—the classic symbolic interactionists and belonging to the 'other' (mental) space—qualitative arguers. Is it really so that a "space of skipped lines" did not exist before Laurel described the life of Louisa May? Or maybe the gap was just too narrow to squeeze through it?

"[…] my epiphany came when I was out on an evaluation contract," confesses Yvonna Lincoln, and reveals that the evil, which she knew little of, suddenly took on a concrete form, and what was worse, it was discovered in a person that was supposed to be an educational model figure ("I began to suspect that in this project, the middle school coach was molesting some of the boys on one of his teams"); suspicions prove to be true, evil emerges from the shadows to display another face of the world in front of the narrator.

"I had a major car accident and I was in a coma for some while," recollects Laurel Richardson, adding that when she recovered she "was not able to do [her] fourth-grade mathematics." Life soaked with extreme experience seeks the return of the fitting existential framework; the narrator feels that with every hurt tissue of the body, concluding: "[…] through the writing, I retaught myself the bases of sociological reasoning. Writing for my life, writing so I would have a life. That was […] epiphany."

Another epiphany of Richardson nourishes itself on the word 'stillness' ("I didn't know how to write. For whom do I write? Whose life can I write? What do I say?"), which throws a bright beam of light on the identity crossroads ("At the same time, I was experiencing the tension between two sides of myself: the scientist and the poet").

The news of the unexpected death of his father put Arthur Bochner at the junction in the world,[15] from where the ends of the horizon are clearly seen ("my days were numbered").

His "serendipitous meeting with Carolyn Ellis" is the epiphany effecting in, as it would be put by Józef Tischner, a significant shift in the inter-human contact (Tischner, 2006b, p. 19) ("I [...] had a serendipitous meeting with Carolyn Ellis in 1990. [...] That [...] was the beginning of a beautiful friendship and the start of our project on ethnographic alternatives").

"My first epiphany occurred when my brother Rex died in 1982 in an airplane crash on his way to visit me. My world was turned upside down," Ellis confides. The emptiness suffered by her is still deepened by the death of Gene Weinstein, and now the loss is already so overwhelming that it makes it impossible to remain in the same place, it is not surprising, therefore, that a subtle suggestion coming from the outside, irrevocably changes the course of events: "My second epiphany came with the death of my partner, Gene Weinstein and the responses I got to my writing [...] about relationship and his dying. [...] to get the varied responses I received was mind-opening and mind-boggling: 'This isn't sociology or research,' 'This threatens the whole sociological enterprise,' and so on. All of it made me more determined to make my case that this was sociology. Norman's response to a paper I did on introspection—that I was being schizophrenic—helped me move from trying to fit into a mainstream so-ciological model to finding my own place on the margins."

Incongruity

Situation, scene, gesture, fraction of a conversation, etc. are epiphanic when *suddenly* and seemingly *without reason* they gain a 'surplus of meaning'; they appear to be something more than they should. Ellis and Bochner's meeting is not just one of a number of meetings on the stage of academic *theatrum mundi*; it gains the importance of an Event.[16] Bochner addresses, but also—literally and figuratively—is addressed.[17] His question induces

---

[15] The phrase 'junction in the world' was created by Józef Tischner (2006c).

[16] "To meet means more than to be aware that the other is present next to me or nearby me. Immersed in the street crowd, I am aware that there are other people nearby me, but that does not mean that I meet them. A meeting is an event," says the philosopher of dialogue (Tischner, 2006c, p. 19).

[17] The meeting-event is mentioned by Bochner and Ellis on a number of occasions present-ing it in various forms, in Ellis's *Revision*, this private-public story is presented as follows: "*Art*: Who is this woman anyway? How could we have been on the same campus for the last six years and never met? Of course, the truth is I was immediately attracted to her

Her answer, and vice versa—Her question is immediately followed by His response, making them The Others for each other.[18]

While Farberman treats the remark of the skipped line in a purely polemical dimension (as an argument supporting the position vested in the discussion), Denzin immediately sees the 'difference' in it that constitutes separate scientific communities.

Lincoln, discovering child abuse committed by a teacher, moves the problem from the realm of legal responsibility into the area of morality; she also sees in it a clear indication for her own explorations reaching ethical foundations of life and work.

Severe disease in Richardson is not limited to somatic losses and injuries carving deep furrows in the psyche; coma, in this case, is a call to open the windows when the door gets locked;[19] it becomes, as it would be put by Arthur Frank, a *call for remoralization*.[20] Disease, as confirmed by the casus of Richardson, is a threat that turned into an opportunity (cf. Frank, 2002, pp. 1-7).

---

passion and energy (to say nothing of her good looks). She seems to be about the same age as me, give or take a few years, and her talk focused on how she had taken care of a partner who died. I couldn't help wondering: Is she available? Unattached? If she is, is she still mourning her loss? Hmmm, I liked her vitality and sense of humor. I thought I ought to at least give it a shot, go up and meet her, feel out the possibilities. So I waited until the crowd around her dissipated and introduced myself. We walked out of the room and into the parking together. *Carolyn*: After talking, we decided to exchange articles we'd published. On Monday, I had a student run some things across campus to Art's mailbox. That day and the next, every time I went to my mailbox, I had another article from Art. Sometimes only an hour apart. Hm, I thought, this is definitely an academic form of flirtation. So I wrote Art a note thanking him for the articles, and added, 'I'm certainly glad we found each other'" (Ellis, 2009, p. 112; form as in the original).

[18] A beautiful passage about 'The Other' can be found in *Ślad i obecność* (*Trace and Presence*) by Barbara Skarga (2002, p. 53), where she writes, "There is an experience in which the whole of our being is focused in one moment, as if through a lens, that very moment makes a rare experience, but one who felt it, knows what it means. Here invades my being the other, who is suddenly met. We meet others all the time, but this one is not in a crowd, he is in front of me, face to face, and at that moment my multi-ecstatic time freezes. The other [...] brings shock, imbalance that has no past behind it and has not opened for the future yet. He shakes me with his presence. I can defend against this otherness, try to reject it, and finally forget it. But the moment of this presence has already changed the course of my life, marked a profound turning point in my existence. The presence of the other may become precious or unbearable. Meeting the oppressor is just as shocking as meeting the lover. And these meetings weave the thread of our lives, marking the moments of presence in our timing."

[19] "When God closes a window, she opens a door"—it is the final sentence of *At the Will of the Body*, by Arthur Frank (2002, p. 156).

[20] "Once the body has known death, it never lives the same again," says Frank in *At the Will of the Body* (p. 16), and elsewhere he writes, "Illness and disability call upon people to become morally engaged because they have everything to lose, but also to gain" (Frank, 2004, p. 177).

Creative (writing) impotence is seen as an identity-based category and not a transitional state of mind.

Finally, the moments when one's beloved leave them (in the case of Bochner, Ellis) go beyond the personal tragedy (combined with a deep human need to experience grief), being converted into the necessity of performing an axiological volt.

'Spirituality'

Epiphany is inseparable from transcendence (Latin: *transcendere*—'to cross'), but depending on the 'ideological base' creating the 'perspective' of our reading out the **epiphanic experience**, it fits either into the absolute being (God) *versus* the limited being (human) relationship, or is exhausted in the plan of man *versus* the external object;[21] the first situation corresponds to theophany, the second—to the secular epiphany. In both cases, the act of 'crossing' refers to reaching 'beyond' the material realm. It is, to look for a visualizing expression, an 'opening up' to such or another **non-ordinariness** (Stephen–Joyce explains Cranly in his further considerations on the clock: "Imagine my glimpses at that clock as the gropings of a spirituals eye which seeks to adjust its vision to an exact focus. The moment the focus is reached the object is epiphanised" (Joyce, 1963, p. 211)). The 'spiritual' attribute of epiphany is also linked to its **timelessness**; epiphany is a 'point,' which means that it shows its strength by throwing a person into a 'momentary void'; in epiphany there *already* is no past, and the *future* has not *yet* arrived; epiphany is 'eternal now.' The meaning of epiphanic eternity now recalls the principle of void by Laozi. "Thirty spokes join the wheel nave/ And make of void and form a pair,/ And a wagon's put to use./ Clay is thrown to shape a vase/ And make of void and form a pair,/ And vessel's put to use./ Door and window vent a room/ And make of void and form a pair,/ And a room is put to use," says the sage (Laozi, 2004, p. 51). Emptiness therefore determines the usefulness of what can fill it in, but what is more—it promotes the launch of source potency, it is a gap, that line skipped by Richardson; thanks to it, almost everything is possible, or at least what was un-think-able is now realized ("That skipped line provoked a give and take in the journals and opened this space that we were in, a space of skipped lines (Denzin)); ("I ended up writing a life-history interview of the unwed mother, 'Louisa May,' [...]. This experience [...] created the space in the discipline and in our world where we could be a community" (Richardson)). A fleeting moment of

---

[21] To read more on understanding transcendence cf. Wciórka (1994, p. 381).

epiphany is defined by its capacious ovary, its modality, subjunctivity (in the sense coined by Turner (cf. Turner, 1995)), combined with the almost unlimited suitability to the tilt 'towards the new.'

Epiphany and 'Biographical Landmarks'

Each **border experience**, and epiphany belongs to that category, is first a presence in itself to use the vocabulary of Barbara Skarga. The suddenness of epiphany, the transcendental nature moves towards the 'raw' being, but—this is very important (!)—is not exhaustive in such being; the 'naked sensation,' as I would put it, is covered by the awareness of it, and this in turn brings to the fore reflexivity. "The moment of reflection," says the philosopher, "and this presence already pulls away and distance is created, even a minimal one" (Skarga, 2002, p. 51),[22] the distance facilitating assessing the gravity of the sensation and its irreplaceability. Epiphany, to remain within the spirit of the impressions of *Kwintet metafizyczny* (*Metaphysical Quintet*), is "my own experience, not for sale or exchange with others, the original experience, from which I draw and that wakes me up, triggering my rise from being submerged in the world, wakes me up to what is [...] my project of being, and if not a project, my own vision of myself and of what surrounds me, the experience confirming something that is, and maybe should be the essence of my existence" (Skarga, 2005, p. 128).

When we highlight this background, it appears advisable to look at epiphany from the perspective—as would proceed Norman Denzin—revealing in it the aspect of '**biographical landmark**.' Epiphany acts then as an indelible marker; it defines the line dividing life into 'before' and 'after' or 'then' and 'now,' *favoring* the reflection of who I am, who I was and who I should be. Epiphany, under certain conditions, appears to be a turning point, because it "wakes up something in me, asks something of me, leads me in a new direction, cancels my previous desires and goals. As if somewhere in the depths of my self a transformation was carried out, sometimes only partial, but already significant, sometimes crucial, penetrating the very fabric of my identity" (ibid., pp. 127-128). Man experiencing a revelation-illumination is someone whose awareness is significantly transformed, in whom a feeling is converted into certainty, or, on the contrary, what was certain, like under a magic spell, becomes immediately questionable, and therefore, requiring retouching.

---

[22] I will track once more the Joyce's interpretation of epiphany, to find—following Dedalus—"the first quality of beauty," which however is sometimes subjected to meticulous judgment. "Beauty," says Stephen, "is declared in a simple sudden synthesis of the faculty which apprehends. What then? Analysis then" (Joyce, 1963, p. 212).

The autobiographical stories of Bochner, Richardson, Ellis and other scientific converts are the visible signs of changes situated between the 'living experience' and the intensive work of consciousness imposed on it, which seeks to turn life upside down almost entirely; in terms of producing new quality and applying it for the purpose of the forming of my 'I' and the world behind it. "If experience has the narrative quality attributed to it here, not only our self-identity but the empirical and moral cosmos in which we are conscious of living is implicit in our multidimensional story. It therefore becomes evident that a conversion or a social revolution that actually transforms consciousness requires a traumatic change in a man's story. The stories within which he has awakened to consciousness must be undermined, and in the identification of his personal story through a new story both the drama of his experience and his style of action must be reoriented. Conversion is reawakening, a second awaking of consciousness. His style must change steps, he must dance to a new rhythm," writes Stephen Crites (2001, p. 43).

At this point, there is a convergence of the views of Crites and the thought of Józef Tischner, situating the issue of consciousness transformation at the junction of the inner and outer world of man. "Transformations of consciousness are an internal reflection of the external human drama," says the author of *Etyka solidarności* (*Ethics of Solidarity*) (Tischner, 2006c, p. 242, see also, Tischner, 1978), basing his argument on the principle of solidarity and egotic de-solidarity; according to it, 'I' can strive in two directions. When 'I' gets entangled around some 'axis values,' it shows some kind of solidarity with itself, and when it removes the value from the field of view, it ceases to be true to itself (Tischner, 2006c, p. 242). But what exactly does it mean to be true to oneself?

Fidelity to oneself—to reach for yet another Tishner's idea albeit close to the discussed one—is a constant man's search for his proper *ethos*, and through it an attempt to reach the essential existential ground, leading me and You to the good that builds us. If a human "discovers where, among which issues, with what kind of people h i s *ethos* is linked, he may 'bear abundant fruit.' If he does not find it, he will live as an alien creature to himself," firmly states the philosopher from Polish Highlands (Tischner, 2006a, pp. 171-172; spelling consistent with the original). His passages on the ways to ethical recognition encourage to looking for such layer—explicitly localized, and who knows, maybe key from the perspective of narrators—of the stories about 'scientific' and 'personal' **conversions**.[23]

---

[23] Conversion (Hebrew: *šub*, Greek: *epistrephein*—change of the way, return, turning back from the way, Greek: *metanoia*—internal conversion), usually associated with the area of religious experience, also has its equivalent in the form of a laic experience, which is understood as "a transition in the inner life to higher degree of excellence"; in this

The theme of seeking personal *ethos* is most clearly evident in the stories told by Ellis and Bochner; they, in my opinion, in a unique way experience the **'desire' to bear witness** on what they experienced. The word 'desire' is attributed here the metaphysical color in the sense presented by Emmanuel Lévinas. "To feel the desire is not the same as to feel the need. The need is always associated with a 'saturation' with this, to which it turns. Intentionality of the need is two-way: the need for food turns to bread, eating bread is the 'satiating' of nutritional values of bread. A similar kind of 'satiating' occurs in the case of looking, listening, many intellectual acts. Anyone who feels the need, also vaguely sees what could satisfy it. [...]

It is different in case of Desire. 'Metaphysical desire,' writes Lévinas, seeks something entirely different, something absolute in its Otherness. Satisfying the desire does not satisfy any hunger, because it reaches those who already satisfied their hungers. The desire is 'unhappiness of the happy.' Even love cannot fall into the category of desire, as it is the hunger and saturation. Desire is more like human kindness and actions arising out of goodness do not saturate goodness, but deepen it further. Anyone who has been good, desires to be good. And so in the desire: what is desired does not fill in the desire, but hollows it even more" (Tischner, 2006b, p. 71, see also, Tischner, 2006c, pp. 28-37).

"There Are Survivors," Carolyn gave that title to her story about Rex's death (cf. Ellis, 1993). This phrase heard on the radio, waking a vain, as soon appeared, hope, included in the told story becomes particularly significant. Ellis turns to those who participated in the flight and actually *survived* the crash, but also to those who, like her, still live even though they have suffered irreparable losses. While grieving she discovers that in order to make the suffering tolerable, she *needs* to start talking about it. She talks and writes *for herself*, while thinking *about others*; her *desire* is to draft a map giving the people who suffer (whoever they are) some clues in the chaos of blurred experience, she *desires* to give evidence that would indicate that life, after all, is more valuable than death.

The autoetnography *There Are Survivors* was printed in 1993, two years before the publication of *Final Negotiations*. Interestingly, the events depicted in the two narratives took place more than a decade earlier. Rex died in 1982, and Gene Weinstein a few months afterwards. I mention this fact considering that silence which lasted for more than a decade is too meaningful to remain interpretively indifferent to it. How then should we understand these ten years of silence?

sense we talk about the "second conversion," evoking—according to the term by Jan Pryszmont (1994, p. 293)—"new personal attitude."

Well, in my opinion, this period of silence is rooted in the bipartite ground—purely *personal*, not to say—intimate and *professional*, associated with the need to develop a vocabulary suitable for the utilization by the scientific collective. Still, it seems that there is a point at which the high-lighted areas converge and overlap.

Sometimes, extreme and traumatic experiences clothe us in silence; I fall silent when the pain is beyond me, when it becomes too great for me to be able to express it, articulate it. This regularity for a long time prevented the Jewish Holocaust survivors from returning to the drama of the war not only in their own memory, but also in verbal recollections (see e.g. Frankl, 1959); the language loses its power also in the face of severe disease (*vide* the case of Laurel Richardson); then the function of speech, of course not directly, is taken over by the body. "Seriously ill people are wounded not just in body but in voice. [...] in the silences between words, the tissues speak" (Frank, 1995, p. xii) (pain forcing to scream?); after all, the context of the disease in general, not necessarily of chronic or terminal nature, works to the disadvantage of 'speaking,' at least at the level of the colloquial use of language. Suffering and pain ruthlessly expose any linguistic non-excellences, which was vividly recognized once by Virginia Woolf (1930, p. 11): "English, which can express the thoughts of Hamlet and the tragedy of Lear, has no words for the shiver and the headache. It has all grown one way. The merest schoolgirl, when she falls in love, has Shakespeare or Keats to speak her mind for her; but let a sufferer try to de-scribe a pain in his head to a doctor and language at once runs dry. There is nothing ready made for him. He is forced to coin words himself, and, taking his pain in one hand, and a lump of pure sound in the other (as per-haps the people of Babel did in the beginning), so to crush them together that a brand new word in the end drops out."

The new word of Virginia Woolf falls within, if treated as a metaphor, the area of Ellis's efforts to find her own voice (i.e.: the reconstruction of the 'I'), which—let me condense the course of events—eventually turned out to be a joint voice of the "wounded storyteller" (Frank, 1995) and a sen-sitive engaged social researcher involved. What did the various stages of reaching this point look like?

Ellis launches from confronting the traumatic experience with the quite rational element, seeing in the therapeutic dimension of writing an Ariadne's thread—a thread capable of leading her out of the labyrinth of the crisis: "I'm just going to sit here and write about what is going on in my life be-cause I'm unable to get through this any other way.' I started to write and was amazed at how writing organized my thoughts and helped me set them aside and move on to what had to be done next. The process was *rationally* therapeutic" (Holman Jones, 2004, §48; form as in the original). In other places we read: "Writing has served to put all the 'little incidents' into a bigger picture of recognizable patterns that contradicts to some extent the sense of 'disorientation and disintegration' that threatened me when I lived this story. These patterns give the impression, quite groundless, of control and rationality, which as [Nancy] Mairs says, 'may save one's sanity even though it can't save one's own or anyone else's life'" (Ellis, 1995, p. 328).[24] "I couldn't stop writing because [...] [writing] might help me and others understand and cope with loss provided day-to-day meaning and continuity. Sometimes I felt I was lighting to make my work meaningful in order to rekindle the deeper meanings of my life—a life now superficially safe, afloat in a rowboat, yet threatened by a sea of continuously challenging waves of response no matter where I placed my oars. In which direction would I go? Or did the strength and unpredictability of the waves make direction a mute concern? Finally, I pulled in my oars and set off on my own adventure, eager to see where it would lead me" (ibid., p. 309).

I have suggested above that Ellis's long silence was due not only to the attempts to cope with the loss as such (personal-intimate dimension of experience), but also the loss which, as directly confirmed by the statements of the author of *Final Negotiations* (cf. ibid., pp. 303-337), in its intention, from some point aspired to take the form of research—ethnographic/ sociological work. The trouble was, however, that neither Carolyn nor anyone from the group of people working with her had the right vocabulary suitable for use in the description of the border experiences without also ripping them off very subtle senses. This problem will become a bit clearer if we approach it from the metalevel of considerations of Ludwik Fleck and, closer to our time, Richard Rorty.

On the occasion of a polemic with a psychiatrist and philosopher of medicine, Tadeusz Bilikiewicz, Fleck says, "It seems that it is less important to consider all ideas and theories, such as embryonic evolution of the

---

[24] Ellis's confession is similar to what Laurel Richardson (2001, p. 33) felt: "After the car accident and coma, I could not find simple words much less remember proper names, dates, titles of papers or journals, or perform statistical tests fourth-grade arithmetic. Although I could not bring into speech what was happening in my head, I found that I could write about it. If I could not find the word I wanted, I could write its first letter or leave a blank space. In writing, the space and the issues were my own, not the maddening questioning of others. Writing allowed me to record little thoughts, to revisit them and fill in the blanks, to piece them together, thought-by-thought. Writing gave me a feeling of control over time and space, and a faith that I would recover. Writing was a method through which I constituted the world and reconstituted myself. Writing became my principle tool through which I learned about my self and the world. I wrote so I would have a life. Writing was and is *how* I come to know" (italics in the original).

eighteenth century, than to carry out an analysis of individual texts, like an analysis of an unknown code. Picturing the content of the view from a past epoch with contemporary expressions is impossible, because the individual terms of this era are incomparable to the present ones. The 'embryo' of the views of the eighteenth century is something very different than the 'embryo' of today's style of embryology. [...] The stylish aura of concepts has changed, and the views change accordingly. Now, we must first examine the aura of concepts, their stylish color, reflecting in the language custom regarding the use of certain words, *particularly their metaphorical use*. Only this opens the door to studying the thought style of the epoch" (Fleck, 2007b, p. 268; italics in the original).

Such revealing feelings of Fleck find support in the findings of Rorty congenial with them. The author of *Consequences of Pragmatism* notes that "scientific breakthroughs are not so much a matter of deciding which of various alternative hypotheses are true, but of finding the right jargon in which to frame hypotheses in the first place" (Rorty, 1982, p. 241). Scientific jargon appears here not as a means by which we are going to describe reality as it really is, but as a means of responding to local, instantaneous demand for given social worlds. There is no 'universal vocabulary of science,' but there are numerous 'particular vocabularies,' "jargons" generating an opportunity of "coping with things"; such a pragmatic approach assumes, in the case of the constitution of any new stream of research (and related to it scientific moral community; the thought collective), that each time there is a need to "hunt for new terminology" and use "casting about for a vocabulary," inventing 'adequate' concepts and containers for them, or their forms of expression. "Vocabularies are useful or useless, good or bad, helpful or misleading, sensitive or coarse, and so on; but they are not 'more objective' or 'less objective' nor more or less 'scientific'" (ibid., p. 203), writes the philosopher of hope. This finding is all the more important since it allows the author to lead us to—equally radical and logical—consequences expressed as follows: "if we get rid of traditional notions of 'objectivity' and 'scientific method' we shall be able to see the social sciences as continuous with literature—as interpreting other people to us, and thus enlarging and deepening our sense of community. [...] The lines between novels, newspaper articles, and sociological research get blurred" (ibid.).

Rorty, following his custom, carries the findings regarding the languages of science onto the level of social sciences to distinguish between two basic current trends in contemporary vocabulary: first, in terms of "descriptions of situations which facilitate their prediction and control," and second, in terms of "descriptions which help one decide what to do" (ibid., p. 197). The dictionary of predicting and control is the **'dictionary of politics'**; the dictionary supporting decision-making is the **'dictionary of ethics.'** I reach for these distinctions, because they allow me to recreate briefly (and thus with some simplification) the trail, followed by Carolyn Ellis and her kin from the qualitative thought collective.

The work 'before the turning point,' established on the basis of initial fieldwork held among fishermen in the Chesapeake Bay, is described by its author in the following words: "My dissertation was organized around 'legitimate' sociological topics—social structure, family, work, and social change. Within this framework, I discussed 'hard' sociological concepts, such as personal attachment, locus of social control, reciprocity, public conformity, civic status, individualism, communitarianism, center, and periphery. Yet, it was difficult to capture the complexity of the lives of the fisher folk using these categories, and I often felt unsure of the distinctions I was forced to make. To me, these theoretical concepts seemed as vague, subjective, and ethereal as emotional experience. I wish now that I had placed more emphasis on how people felt, which was my primary interest" (Ellis, 1995, p. 6).

Through the last sentence Carolyn clearly shows her preference for choosing *the dictionary of ethics*, since, instead of turning her face *towards the stage* (Tischner (2006c, p. 7) would call it the "attitude of intentional objectivity"), she directs it *towards man*; going in for acting *between us*. "I also wish now that I had been more present in my writing about the fishing communities. Mostly, I describe 'them,' the fisher folk, interacting with each other, as though I am off in the corner, invisible. In reality, most of what I learned came through my interactions with the people, especially their reactions to me. [...] I anguished over speaking in the first person, having been told it was 'unprofessional' and that readers would then conclude I had not been neutral and distant" (Ellis, 1995, p. 6).

Accents postulating the presence of me-researcher *with* them—the presence that should be updated by an ethnographic/sociological text—appear in the final stage of the consciousness work coming after 'the turning point' experiences and are associated with a strong imperative of soaking social sciences with humanistic element in general. This is not to be an individualized path, followed by a single wanderer-conquistador. On the contrary—Ellis with all her strengths wishes to create a "collective project that seeks to humanize sociology, create a space for experimental texts, and encourage writing stories that have meaning and make difference in people's lives" (ibid., p. 336). That project began in a place of dramatic existential pain, and then was transferred into a public place, occupied by a man sunk—as Lévinas would say—in poverty, someone earnestly seeking his *axis mundi*. In preparing the original version of *Final Negotiations*, the author feels great dilemmas: "I knew I did not want to write a traditional academic book and that I wanted to write a book that would speak therapeutically to a mass audience and sociologically to an academic one. But, I feared there were too many stories to tell to too many people [...]. What kind of book would I write? Would I tell my story, or use it as grist for sociological analysis? Would it be a scholarly or trade book? Would I write to social scientists or a mass audience? If social scientists, would it be all sociologists or some sympathetic section of them, such as symbolic interactionists, those to whom I had addressed my papers? Would it be a self-help book or a contribution to the sociological literature? These issues arose as I was going through intense grief. The losses of Gene (1985) and my brother (1982) were renewed when first my dog of fourteen years died (1986), and then my father passed away (1987). Any questioning reaction to my work presented more threat of loss—loss of identity—and I often felt defensive and judgmental in response to criticism. Similar to irresolvable grief that often results from ambiguous feelings felt toward a loved one who dies, negative criticism coupled with indications that my work was important made it more difficult for me to resolve how to accomplish this project" (ibid., pp. 308-309). Ellis's struggles are not wasted, they are forged into a "discourse [that] connects human experience as live to research on emotions and intimate bonds, permitting **my 'heart and head [to] go hand and hand.'** [...] As this discourse becomes a part of sociological perspective, my life and work are coming together in meaningful ways. The result is a sociology that connects life experience to the pursuit of knowledge. This view makes the activity of doing sociology **personally meaningful** [it is intertwined with the personal *ethos*—M.K.] and anticipates for sociology what [Victor] Turner once said about anthropology, that it can become 'something more than cognitive game played in our heads and inscribed in—let's face it— somewhat tedious journals'"] (ibid., p. 335; bold added by M.K.).

03

"I must take charge of my life/story" (ibid., p. 333), categorically rules Carolyn Ellis, clearly defining in this way the direction of her new path. Arthur Bochner faces a similar challenge; on the day of his father's death he enters a border situation and this always requires from a man posing critical questions. As Józef Tischner (2006a, pp. 186-187) says, "In the 'border situation' a man always brushes against the 'border' of his human existence. At that point, we either lose ourselves or recover oneselves. Because each time, we build and destroy something in us. A man shows his true nature in the 'border situation.'

The basic example of a 'border situation' is experiencing responsibility. I am what I am. I am responsible for myself. Mine is the merit, mine is the fault. I have the courage to look into my own interior and accept myself without any dodging.

An important hallmark of a person is the ability to take responsibility. Nothing demonstrates better the maturity of a human being. Man was thrown into the world both real and ideal, and is responsible for the extent to which these two worlds meet each other."

The author of *It's About Time*, upon his father's death, experiences a collision of two worlds—the world of hard science, which he previously dabbled in and the world of mundane, tangible human feelings assuming deep meaning after what had happened. Bochner knows (or only dimly anticipates) that the choice he makes will determine his future life. He can lose himself, he can save himself…

"I felt dazed and confused, like a boxer startled by the first powerful blow from a stronger opponent. Stunned by the punch, he hears competing voices, one inside his head whispering, 'Ignore the pain, stay with the game plan,' the other calling from the site of his body's pain and injury, rejecting the authority of consciousness over bodily experience.

A voice inside my head said, 'Get home to Tampa as quickly as possible. Mother will need you. She'll expect you to take control, help arrange the funeral, and keep the family from falling apart.' Suddenly, the three papers I was to present at the convention had little significance. However, I was too responsible to miss sessions without forewarning. I should contact the chair of each program, get someone to substitute if possible, give other participants to prepare for my absence.

But a second voice kept intruding on my own thoughts. I felt dizzy and lightheaded, as if I were teetering on the edge of a dangerous cliff. As I wiped away the tears trickling down my face and felt the flood of anxiety swirling through my stomach, I was terrified to realize that I couldn't shut down what I was feeling by an act of willful control. My father's death was not just another event to be organized, experienced, and filed away. It wasn't only my plans for the weekend that had been interrupted, but something much bigger" (Bochner, 1997, p. 419).

When Art—overwhelmed with the absoluteness of the experience—is not strong enough to lift up the phone receiver and tell his friends about the situation, something unexpected happens, there appear signs of great inner transformation that carries heavy consequences both for Bochner-man and Bochner-scientist: "As I sat in the corner and watched Herb organizing my affairs, I recalled the times I had tried to talk to him about interest on death and dying. These conversations never got very far. Herb resisted invitations to delve deeper and I usually felt disappointed that we couldn't connect on this topic. Now, I was beginning to understand why these conversations have been so frustrating and superficial. At the time, Herb's parents were dead; mine were alive. For Herb, death had been personalized, for me, it was academic. Under these circumstances, what did we really have to talk about? How could we possibly speak the same language? As a result of my father's death, I had passed into another dimension, one that was missing when Herb and I had tried previously to converse about death. We still weren't talking about death or loss, but when Herb looked at me from across the room, I felt the kind of communion that can only occur when two people are woven into the same fabric of experience.

On that long plane ride home, I realized as never before that I was a human being. It sounds strange to say that, I know, but I believe it is true. My father's sudden death forced me to grasp the significance of how contingent, limited, and relative human experience can be. Most of us realize that fear of death lingers behind the absorbing details of our everyday lives, but we keep our fear sedated because we sense it could infect us if we let go. When our lives are interrupted by the reality of death, our immunity is weakened. Then, if we allow it, we can drop the canopy of dishonesty covering the brute fact that we don't really control our lives. [...]

My father's death made it possible, even necessary, for me to see the consequences of splitting the academic self from the personal self in a new light. At my university, or at conferences, I normally move in and out of analytical or conceptual frames without experiencing anything akin to an experiential shock or epiphany. But when my father died while I was attending a national communication convention, two worlds within **me** collided, and I was stunned to learn how tame the academic world is in comparison to the wilderness of lived experience. [...]

Now, the academic man in me stood face-to-face with the ordinary man. What did they have to say to each other? Could they get in touch with each other? Integrate? Harmonize?" (ibid., pp. 420-421; bold in the original).

The confession of Arthur Bochner, expressed in the form of an internal dialogue, largely directed towards himself, imperceptibly turns into a call and challenge,[25] gaining Tischner's spirit of working on exploring and discovering *ethos*: "My personal struggle after my father's death was

---

[25] I use the division proposed by Małgorzata Czermińska (2000).

not a scientific crisis but a moral one," says Art and adds, "and the moral questions that were raised cast a long shadow over both my personal and my academic lives. I needed to take measure of my own life and of my father's too. How were the different parts of my life connected? What values shaped the life I wanted to live? Why would my academic life be if I could bring those values into play? What would it feel like?" (ibid., 423). Running away from objectivism, within which (no matter what the reasons were) Bochner operated before, after the traumatic events he goes on a journey with a one-way ticket: referring to, *inter alia*, the views of Stanley Milgram, Thomas S. Kuhn, Ronald D. Laing, Keneth J. Gergen, and first and foremost Richard Rorty that are like drops of water hollowing out the rock called the scientistic image of the world, he demands from himself a radical change of the ideological optics from one mounted on *a quest to learn* to one involving *the desire to act justly*; hence there arises the postulate regarding the need to move away from the pure scientific theory, to implement something that Art gives the name of **social theory**. While the first is a weave of abstract concepts and terms, considered to be aloof to life and free from values, the second assumes as its goal vivid participation in building a reality based on the idea of social and communication activity: "In the world of social theory, we are less concerned about representation and more concerned about communication. We give up the illusions of transcendental observation in favor of the possibilities of dialogue and collaboration. Social theory works the spaces between history and destiny. The social world is understood as a world of connection, contact, and relationship. It also is a world where consequences, values, politics, and moral dilemmas are abundant and central" (ibid., p. 435).

Conclusion

Finally, I propose to draw attention to the issue, which will serve as a buckle closing the output of my thoughts. Emile Cioran in his book, *Tears and Saints*, ventured to say that "All conversions are sudden but they take years germinating underground. [...] Divine revelations break out after a long period of incubation" (Cioran, 2003, p. 96). Please, remember, Dear Reader, Stephen's dialogue with Cranly... Ballast Office clock, seen every day by Dedalus hiking in the streets of Dublin, once finally becomes an epiphany; but... before it happened... many gestures were made (Stephen: "I will pass it time after time, allude to it, catch a glimpse of it"), constituting "the gropings of a spirituals eye which seeks to adjust its vision to an exact focus. The moment the focus is reached the object is epiphanised" (Joyce, 1963, p. 211).

Therefore: epiphany is not hung in a void, not always and not everywhere.

"Mocking my fears and hopes, flashbacks of live TV footage of passengers from my brother's plane floundering in the Potomac River were interrupted in real life by Gene choking and yelling for me to untangle his oxygen hose. Suddenly, the scientifically respectable survey of jealousy I was working on seemed insignificant. Instead, I wanted to understand and cope with the intense emotion I felt about the sudden loss of my brother and the excruciating pain I experienced as Gene deteriorated," says Carolyn Ellis (1995, p. 8), poignantly describing the moment of experiencing the awareness change. It was not, however, as it turns out, a radical change founded on the ashes of the old world, but rather a bold, full opening of the already repealed door. In the *Introduction* to *Final Negotiations* (entitled *Beginning*), there are micro-stories approximating the academic life meanders of the researcher 'before the transformation.' The narrator presents there herself as a character thoroughly ambivalent, torn between the desire to meet the serious scientific challenges (requiring adherence to the rigor of objectivism and distancing oneself emotionally from the subject of study) and pursuing close relationships with people. The very first fieldwork, i.e. the one carried out in the Chesapeake Bay, evoke in the researcher a feeling of identity contradiction, deepening day by day: "I often experienced conflict between remaining uninvolved and distant, as I had been trained, and participating fully; between recording only my 'objective' observations of fisher folks' actions and speech and noting my sense of their emotional lives, a process that required my engagement. Often distance won out over involvement because of my concern about meeting the requirements I had learned for being a neutral social scientist.

When I returned to the university to write my dissertation, I struggled with the constraints of detached social-science prose and the demand to write in an authoritative and uninvolved voice. Though I worked hard to follow these principles, professors I admired still reprimanded me for having 'gone native' and for being too sympathetic toward my subjects" (ibid., p. 6).

In the following paragraphs, Carolyn writes, "Even during this research, however, I was drawn to stories for conveying lived experience and insisted on inserting vignettes showing specific incidents. In these stories, I could occasionally be present, though I rarely got to speak and almost never got to feel. But I knew, even then, that I wanted readers to hear the participants' voices and see them acting. The vignettes breathed life into my more passive telling and categorizing of the fisher folk" (ibid., p. 7).

Seen in this context Ellis's epiphany seems to crown a specific stage of the struggle of the 'I' in order to become a leaven of their next revelation; it is therefore a unique point, however inscribing itself in a very wide arc of life.

Similar motifs, if not identical in their meaning in relation to those tracked in the story of Ellis, can be found in the biographies of Bochner and Richardson and in texts corresponding with them. The father's death induced the author of *It's About Time* to make a bold move, but was it the real beginning of the transformation? In September 2010, during my internship at the Department of Communication at the University of South Florida (the *alma mater* university of Bochner and Ellis), I asked Art about the beginnings of his interest in qualitative methodology. To my surprise, he did not mention his father's death, but at once referred to other tragic events. He told me about the mental illness of his first wife and trying to cope with that situation, and also about the thoughts aroused in him by reading books; first, in 1974, *Pathway to Madness* by Jules Henry (1965), and later Robert Coles's *The Call of Stories* (1989).

What is extremely intriguing is also the background against which Laurel puts herself: "I'm sitting here and deciding whether to start my qualitative experience at my birth. I think the fact I've been a marginal person with a foot in two different culture[26] from my birth on did construct me as a sociologist and later as a qualitative researcher. It's a gift. I've been fortunate to be born into two different cultures. What moved me forward was the capstone course I took as an undergraduate at the University of Chicago. That course was called The Organization, Methods, and Principles of the Sciences. In that course, we dealt with intuitive work, deductive work, induction, reduction; and it wasn't like there was only one way to do science. Rather, there were multiple ways by which you might come to know something" (Ellis et. al., 2007, p. 236; form as in the original).

At the beginning of my article I suggested that a qualitative revolution would not have happened but for the emergence of liminal individuals, fitting the 'spirit of the times.' It is an appropriate time for a brief development of this thought.

In *Dramas, Fields, and Metaphors*, a collection of essays explaining and exemplifying the theory of 'societal change,' its founder, Victor Turner, pointed out the outstanding function played in the process of innovative cultural change by "*conscious* human agents" (cf. Turner, 1978, pp. 16-17; italics in the original). These "threshold thinkers" with a non-conformist attitude towards the ambient world, "the unacknowledged legislators of mankind," seeing farther, harder and deeper than others, "prophets"—

---

[26] Laurel Richardson is of Jewish origin (footnote—M.K.).

these names are used by the eminent anthropologist (ibid., p. 28)—have a tendency to cross rigid conventions, to test the valid standards and rules in order to focus on making maximum use of "the powers that slumber within man." "In liminality," says Turner, "resides the germ not only of religious *askesis*, discipline, and mysticism, but also of philosophy, and pure science" (ibid., p. 242; form as in the original). Both tasks are completed by those who became 'men apart' belonging—by their own choice, or an external indication—to the social, existential and mental (imaginary) border zones marginal, and peripheral to the areas—it is necessary to add—that enable those residing in them perception of the reality in the perspective unattainable for people colonizing cultural centers. To the group of "liminal thinkers," Turner includes, *inter alia*, those feeling the spirit of change, post-renaissance writers, artists and philosophers (ibid., p. 28), and I do not hesitate to place beside them the creators of the qualitative thought collective. I was encouraged to make such a move, coincidentally, by Ludwik Fleck [*sic!*], carefully, not to say—liminally read Ludwik Fleck. Revealing the backstage of scientific discoveries, the author of *Genesis and Development of a Scientific Fact* writes that they are carried out by the appearance of intellectual excitement and tendency to change in the era of a balance that gave rise to "a chaos of contradictory, alternate pictures. The picture, fixed up now, disintegrates into blots which arrange themselves into different, contradictory shapes. From other fields, previously separated or neglected, some motives are added; historic connections, almost accidental, various intellectual relics, often also the so-called errors, mistakes and misunderstandings for their part add other motives. **At this creative moment there becomes embodied in one or more investigators the mental past and present of the given thought-collective. All physical and mental fathers are with them, all friends and enemies. Each of these factors pulls to its side, pushes or inhabits. Hence the flickering chaos. It depends on the intensity of feeling of the investigator whether the fact, whether the new shape will appear to him within the chaos as a symbolic vivid vision, or else as a weak hint of a resistance which inhabits the free, almost discretional choice between alternate pictures**" (Fleck, 1986c [1935], pp. 76-77; bold added by M.K.).

The pattern of the emergence process of a 'form' serving as the operator of a given thought collective, perfectly fits the atmosphere penetrating the stories of Denzin, Richardson, Bochner and Ellis. Their performances, provocative speeches, suffering insults from orthodox scholars, general anxiety of the research soul, strenuous search for new inspirations meant to provide the theoretical and methodological points of reference, and finally creation of their own language tools that allow for the construction of the dictionary congruent with *our* beliefs, is nothing else but Fleck's chaos, causing mood tension among the group of converts, or—as Turner would rather call it—a liminal state, desired by *conscious* human subjects that have chosen to follow their path.

Is there, in the midst of all that I have mentioned above, also the place for epiphanic experiences and transformations of consciousness lurking behind them? The unconvinced are asked to be patient.

References

Beja, M. (1971). *Epiphany in the Modern Novel.* Seattle: University of Washington Press.

Bochner, A.P. (1997). It's About Time: Narrative and the Divided Self. *Qualitative Inquiry, 3,* 418-438.

Cioran, E.M. (1995). *Tears and Saints.* (I. Zarifopol-Johnston, Trans.). Chicago: University of Chicago Press.

Conrad, J. (1921). Preface. In J. Conrad, *The Nigger of the Narcissus: The Works of Joseph Conrad* (pp. vii-xiii). London: William Heinemann.

Coles, R. (1989). *The Call of Stories: Teaching and the Moral Imagination.* Boston: Princeton University Press.

Crites, S. (2001). The Narrative Quality of Experience. In L.P. Hinchnam & S.K. Hinchnam (Eds.), *Memory, Identity, Community: The Idea of Narrative in the Human Sciences* (pp. 26-50). Albany: Suny Press.

Czermińska, M. (2000). *Autobiograficzny trójkąt: Świadectwo, wyznanie, wyzwanie.* Kraków: Towarzystwo Autorów i Wydawców Prac Naukowych Universitas.

Ellis, C. (2009). *Revision: Autoethnographic Reflections on Life and Work.* Walnut Creek: Left Coast Press.

Ellis, C. (2004). *The Ethnographic I: A Methodological Novel about Autoethnography.* Walnut Creek: AltaMira Press.

Ellis, C. (1995). *Final Negotiations: A Story of Love, Loss, and Chronic Illness*: Philadelphia: Tample University Press.

Ellis, C. (1993). "There Are Survivors": Telling a Story of Sudden Death. *Sociological Quarterly, 34,* 711-730.

Ellis, C. (1991). Sociological Introspection and Emotional Experience. *Symbolic Interaction, 14,* 23-50.

Ellis, C., Bochner, A.P., Denzin, N.K., Lincoln, Y.S., Morse, J.M., Pelias, R.J., & Richardson, L. (2007). *Coda: Talking and Thinking about Qualitative Research.* In N.K. Denzin & M.D. Giardina (Eds.), *Ethical Futures in Qualitative Research: Decolonizing the Politics of Knowledge* (pp. 229-267). Oxford: Berg Publishers.

Denzin, N.K. (1989). *Interpretive Interactionism.* Newbury Park – London – New Delhi: Sage.

Fleck, L. (2007a). *Style myślowe i fakty: Artykuły i świadectwa.* S. Werner, C. Zittl, & F. Schmaltz (Eds.). Warszawa: Wydawnictwo Instytutu Filozofii i Socjologii Polskiej Akademii Nauk.

Fleck, L. (2007b). Nauka a środowisko. In S. Werner, C. Zittl, & F. Schmaltz (Eds.), *Style myślowe i fakty: Artykuły i świadectwa* (pp. 264-270). Warszawa: Wydawnictwo Instytutu Filozofii i Socjologii Polskiej Akademii Nauk.

Fleck, L. (1986a). *Cognition and Fact: Materials on Ludwik Fleck.* R.S. Cohen & T. Schnelle (Eds.). Series: Boston Studies in the Philosophy of Science, Vol. 87. Dordrecht – Boston – Lancaster – Tokyo: D. Reidel Publishing Company.

Fleck, L. (1986b [1947]). To Look, To See, To Know. In R.S. Cohen & T. Schnelle (Eds.), *Cognition and Fact: Materials on Ludwik Fleck* (pp.129-152). Series: Boston Studies in the Philosophy of Science, Vol. 87. Dordrecht – Boston – Lancaster – Tokyo: D. Reidel Publishing Company.

Fleck, L. (1986c [1935]). Scientific Observation and Perception in General. In R.S. Cohen & T. Schnelle (Eds.), *Cognition and Fact: Materials on Ludwik Fleck* (pp. 59-78). Series: Boston Studies in the Philosophy of Science, Vol. 87. Dordrecht – Boston – Lancaster – Tokyo: D. Reidel Publishing Company.

Fleck, L. (1979 [1935]). *Genesis and Development of a Scientific Fact.* (T.J. Trenn & R.K. Merton, Eds.; F. Bradley & T.J. Trenn, Trans.). Chicago – London: University of Chicago Press.

Frank, A.W. (2004). Moral Non-Fiction: Life Writing and Children's Disability. In P.J. Eakin (Ed.), *The Ethics of Life Writing* (pp. 174-194). Ithaca – London: Cornell University Press.

Frank, A.W. (2002). *At the Will of the Body: Reflections on Illness.* Boston – New York: Houghton Mifflin Company.

Frank, A.W. (1995). *The Wounded Storyteller: Body, Illness, and Ethics.* Chicago – London: University of Chicago Press.

Frankl, V.E. (1959). *Man's Search for Meaning.* New York: Washington Square Press.

Garrow, D.J. (2004). *Bearing the Cross: Martin Luther King, Jr., and the Southern Christian Leadership Conference.* New York: William Morrow & Company.

Gornat, T. (2006). *"A Chemistry of Stars"*: *Epiphany, Openness and Ambiguity in the Works of James Joyce.* Opole: Wydawnictwo Uniwersytetu Opolskiego.

Henry, J. (1965). *Pathways to Madness.* New York: Random House.

Holman Jones, S. (2004). Building Connections in Qualitative Research: Carolyn Ellis and Art Bochner in Conversation with Stacy Holman Jones [113 paragraphs]. *Forum Qualitative Sozialforschung/Forum: Qualitative Social Research, 5* (3), Art. 28, http://www.qualitative-research.net/fqs-texte/3-04/04-3-28-e.htm [last accessed: June 15, 2007].

Joyce, J. (1963). *Stephen Hero.* New York: New Directions Publishing Corporation.

Kafar, M. (2010). O przełomie autoetnograficznym w humanistyce: W stronę nowego paradygmatu. In B. Płonka-Syroka & M. Skrzypek (Eds.), *Doświadczenie choroby w perspektywie badań interdyscyplinarnych* (pp. 335-352). Wrocław: Akademia Medyczna im. Piastów Śląskich we Wrocławiu.

Kuhn, T.S. (1996). *The Structure of Scientific Revolutions.* Chicago – London: University of Chicago Press.

Kuhn, T.S. (1977). *The Essential Tension: Selected Studies in Scientific Tradition and Change.* Chicago – London: University of Chicago Press.

Langkammer, H. (1990). *Słownik biblijny.* Katowice: Księgarnia Św. Jacka.

Laozi (2004). *Dao De Jing: The Book of the Way.* (M. Roberts, Trans.). Berkeley – Los Angeles – London: University of California Press.

Milton, J. (2003 [1667]). *Paradise Lost.* London: The Folio Society.

Nichols, A. (1987). *The Poetics of Epiphany: Nineteenth-Century Origins of the Modern Literary Moment.* Tuscaloosa – London: University of Alabama Press.

Pryszmont, J. (1994). Nawrócenie [entry]. In Z. Pawlak (Ed.), *Katolicyzm A–Z* (pp. 293-294). Poznań: Księgarnia Św. Wojciecha.

Richardson, L. (2001). Getting Personal: Writing-stories. *Qualitative Studies in Education, 14,* 33-38.

Richardson, L. (1993). Poetic Representation, Ethnographic Presentation and Transgressive Validity: The Case of the Skipped Line. *The Sociological Quarterly, 34,* 695-710.

Richardson, L. (1977). *The Dynamics of Sex and Gender: A Sociological Perspective.* Chicago: Rand McNally.

Rorty, R. (1982). Method, Social Science, and Social Hope. In R. Rorty, *Consequences of Pragmatism (Essays: 1972-1980)* (pp. 191-210). Minneapolis: University of Minnesota Press.

Skarga, B. (2005). *Kwintet metafizyczny*. Kraków: Towarzystwo Autorów i Wydawców Prac Naukowych Universitas.

Skarga, B. (2002). *Ślad i obecność*. Warszawa: Wydawnictwo Naukowe PWN.

Tischner, J. (2005). *Świat ludzkiej nadziei*. Kraków: Wydawnictwo Znak.

Tischner, J. (2006a). Etyka wartości i nadziei. In J. Tischner (A. Bobko, Ed.), *O człowieku: Wybór pism filozoficznych* (pp.169-186). Wrocław – Warszawa – Kraków: Zakład Narodowy im. Ossolińskich.

Tischner, J. (2006b). Emmanuel Lévinas. In J. Tischner (A. Bobko, Ed.), *O człowieku: Wybór pism filozoficznych* (pp. 65-80). Wrocław – Warszawa – Kraków: Zakład Narodowy im. Ossolińskich.

Tischner, J. (2006c). *Filozofia dramatu*. Kraków: Wydawnictwo Znak.

Tischner, J. (1978). Solidaryzacja i problem ewolucji świadomości. In W. Stróżewski (Ed.), *Studia z teorii poznania i filozofii wartości* (pp. 91-102). Wrocław: Wydawnictwo Ossolineum.

Tischner, J., & Kłoczowski, J.A. (2001). *Wobec wartości*. Poznań: Wydawnictwo Polskiej Prowincji Dominikanów.

Turner, V. (1995). *The Ritual Process: Structure and Anti-Structure*. New York: Aldine de Gruyter.

Turner, V. (1978). *Dramas, Fields, and Metaphors: Symbolic Action in Human Society*, Ithaca – London: Cornell University Press.

Wciórka, L. (1994). Transcendencja [entry]. In Z. Pawlak (Ed.), *Katolicyzm A–Z* (p. 381). Poznań: Księgarnia Św. Wojciecha.

Wielki słownik wyrazów obcych PWN (2010). Warszawa: Wydawnictwo Naukowe PWN.

Woolf, V. (1930). *On Being Ill*. London: Hogarth Press.

Woolf, V. (1960). *To the Lighthouse*. London: Hogarth Press.

Chapter Three

LIVING IN RELATIONS

BIOGRAPHIES OF SCIENTISTS
IN THE CONTEXT OF THE ACTOR-NETWORK THEORY

by Michał Wróblewski

Introduction

It is a cliché to say that existence is a complex and intricate matter. If we consider our lives, we become aware of the multiplicity of issues that need to be dealt with in order to achieve a given goal. At the same time, we also distinguish a number of factors that either restrict or facilitate our actions. Each one of us belongs to a particular institutional field, occupying certain space, meeting certain people, and following certain patterns of behavior. Concurrently, each of us also operates within a milieu comprised of objects, values, and ideas, all of which are used to carry out our duties while allowing for creative activities that enrich this intricate structure with new elements.

Nevertheless, it seems that we tend to neglect these complexities when studying the lives of scientists. Why is that so? First of all, it stems from the fact that we enter a space that has been perceived in our culture as special, and within which a scientist has been enjoying a privileged status. To understand the nature of these relations, it may be useful to refer to the concept of an epistemological relation (Zybertowicz, 1995, pp. 73-74), which serves as the basis for the realist paradigm of practicing science conditioning its distinguished status. It consists of: a self-transparent knowing subject (one that is able to abstract from its pre-judgments), a language (whose meanings are rooted in the rational thought), and an object (a reality that can be directly known).

Adopting an epistemological relation, which carries the baggage of naive realism, results in picturing scientists' actions as walking a straightforward way to meet the goal, be it a breakthrough discovery or a revolutionary invention. In that way, a scientist's life becomes a project, with him/her playing the lead role and being the privileged entity with an access to an objective reality. Thus, unlike the so-called regular people, scientists are able to achieve their objectives by virtue of the three elements making up an epistemological relation, and not through dealing with other individuals, particular objects, values, ideas, and the whole complex tissue of everyday life. It would seem that some people on the planet are cut from a different cloth—floating above the worldly matters and capable of ground-breaking discoveries. A scientist is not limited by his/her prejudgments, the society he/she lives in, or the culture he/she comes from, since ultimately these factors do not influence the content of the produced knowledge. Relativity and contingency of the context in which scientists operate are perceived only as obstructions on the way to cognition. A truly outstanding scientist is able to abstract from the socio-cultural noise that surrounds him/her in order to focus his/her efforts exclusively on uncovering, in an objective manner, the regularities concealed in nature.

Nonetheless, if we stray from the paths set out by the traditional philosophy of science[1] as well as the classical sociology of knowledge[2] and focus on the life stories of scientists,[3] it can be observed that the epistemological relation is to a large extent a factor that mythologizes scientific investigation. Biographies of great inventors along with the stories of their

---

[1]   Insistence on granting a privileged status to scientific knowledge has been most pronounced in philosophical debates on the criterion for demarcation, whose objective was to delimit an objective area that for scientists constitutes the main object of interest. The line of thought striving to separate 'knowledge' from 'non-knowledge' constitutes, in one way or another, the unalterable core of science, which is not subject to external influence (social, cultural, etc.).

[2]   An example of the traditional approach is the sociology of knowledge interpreted by Max Scheler. As indicated by the commentators of his writings, "assuming that social determinants of knowledge did not prove its epistemological validity, he in a away accepted in advance the fundamental compatibility between the principles of the sociology of knowledge he had been developing and the epistemological phenomenological program, at the same time regarding them to be logically primary to social-cognitive claims" (Czerniak & Węgrzecki, 1990, p. XXI). For a division into the classic and nonclassic sociology of knowledge see Zybertowicz (1995, pp. 18-25).

[3]   For stylistic reasons, the phrases 'lives of scientists' and 'biographies of scientists' will be used interchangeably.

achievements prove particularly helpful in this regard. Browsing through their voluminous stories makes one realize that the accounts capture in their entirety the heterogeneity and complexity of the various spheres of social life, both of which our culture has removed from the field of science. In other words, a thorough overview of the biographies of scientists creates an opportunity for demythologizing the figure of scientist while making away with false beliefs concerning how a particular discovery or invention was made.

In the present paper, I have outlined some lives of scientists as seen through their complexities. I have rejected the notion concerning the teleological meaning of a given cognitive activity (a self-transparent subject-scientist sets his/her own objective, chooses the means for its fulfillment, and, by virtue of being innately rational, strives to achieve it) as well as the individualistic idea of subjectivity (a subject-scientist is the only acting element that can lead to the achievement of the aim). My analysis draws on the Actor-Network Theory (ANT), which has been widely discussed also in Poland. Dating back to the end of the 1970s, the approach has been developed predominantly by Michel Callon, John Law, and Bruno Latour. My aim consists not so much in discussing this enormously interesting theoretical proposition,[4] but rather in demonstrating why adopting ANT can yield numerous interesting and cognitively productive interpretations, which can be useful to all researchers of scientists' biographies.

Initially, ANT was used to study the dynamics of modern science. The Actor-Network Theory stems from the so-called laboratory ethnography, whose goal was to reach to the practical actions of researchers from all fields of science (Abriszewski, 2010). This was to be done using empirical studies similar to those conducted by anthropologists who collect their material through field studies. As a consequence, ANT is suitable for analyzing biographies of researchers from the natural sciences and humanities alike.

I am going to try to answer the following questions concerning biographies of researchers: what shapes the life of an individual in such a way that it follows a particular scenario? What affects the individual? Which factors determine the individual's behavior, and which are subject to the individual's influence? What exactly defines the context of the individual's activity? What makes the individual achieve his/her objective (make a breakthrough discovery, elaborate a pioneering invention)? What issues does he/she face? What kind of negotiations have bearing on the course of his/her life and the ability to reach a goal?

---

[4] This has been successfully done by others: cf. Sojak (2004, pp. 233-266), Bińczyk (2007, pp. 189-250), Abriszewski (2008a).

Network, or the Context

A study of a scientist's life often begins with a reconstruction of the context in which he/she came to live and work. The life of Copernicus is described in the context of the Neo-Platonic philosophy which regards the Sun as a metaphor of God or the great geographical discoveries (Kuhn, 1957). Plato's work becomes comprehensible only in a context that takes into account the cultural process of shifting from oral to written communication (Havelock, 1963).

ANT steers clear from a simple understanding of the term 'context,'[5] claiming that the majority of researchers accept it as a given, unproblematic term that does not call for an in-depth understanding. Although it continues to be examined, its meaning is subject to extensive modifications. Latour and others interpret context using the metaphor of a 'network.' What is a network? To begin with, it is a set of *acting actors* who influence one another in the course of organizing the string of events. Moreover, is it also *a system of heterogeneous elements* connected with specific relations. Networks are produced as a result of *translations*. Translations involve attaching new elements (actors) to the network in such a way that the entire network consequently undergoes a transformation, while the essence of the actor itself also becomes modified. The context/network is never static; it perpetually continues to *reconstitute* itself.

The Actor-Network Theory involves various types of translation. A translation takes place each time a complex system is reduced with the aim of exercising control over heterogeneous factors and placing them in the media in which they can be addressed. A good example of a translation is drawing a map (Latour, 1999a, pp. 24-79). Cartographic skills make it possible to 'squeeze' a very large and three-dimensional space on a two-dimensional piece of paper that fits on a desk. Hence, a map stands for translating a complex element into an uncomplicated one in such a way that it generates a simple chain of relations between the person looking at the map and the territory itself. As a consequence, a man tracing a route to a destination refers to the real space, but does so through the medium —a piece of paper.

Obviously, the situation can be easily made more complicated. Drawing an adequate map requires adopting the appropriate system of measurement determining e.g. the scale, distance, and heights. Thus, it is yet

---

5   "I have never understood the fascination with the context. A frame can embellish a painting, direct the viewer's attention, or increase the value, but it does not add anything to the work itself. The frame, or context, is a sum of factors, with no bearing on the data, as is commonly known," claims Latour (2010, p. 207).

another example of translation. The real space becomes translated into symbolic meters, ratios, and segments. Various relations are established between the traveler and the space in which he/she operates. The traveler is required not only to be able to read, but also to comprehend the symbols used in geography and cartography. Would it be possible to go even further and ask why the symbols look the way they do? What made it possible to generate the terms which describe spatial relations in an adequate manner? Is the meter a result of a given social context? Who created the scale and for what purpose? Following one translation after another produces a network of dependencies, where material, symbolic, academic, social, and even psychological and economic elements will be in a state of constant flux. This is what ANT considers to be a network.

Adopting the ANT point of view means that subject's actions need to be placed within the translation-shaped network of relations. In other words, it is possible to understand quite fully the actions of a particular person only after all relevant factors have been taken into consideration. In addition, the relational context is so closely 'interwoven' with the individuals situated in it that from the methodological perspective there is no point in separating these two entities, as it would inevitably distort the larger picture. Human identity is in fact constituted as a consequence of a series of translations between the networks it is related to. Man is a collection of heterogeneous materials determining who he is. John Law summarizes the idea as follows: "If you took away my computer, my colleagues, my office, my books, my desk, my telephone I wouldn't be a sociologist writing papers, delivering lectures, and producing 'knowledge.' I'd be something quite other" (Law, 1992, p. 4). Who I am is determined by a number of factors which I deal with, but cannot control. As an acting subject I possess a range of individual skills (such as education or manual talents). Although I can make use of them in order to achieve a particular goal, my actions are always filtered through an exterior network of relations. I can be a Ph.D. student not only because my innate intelligence proved useful in passing the exams, but predominantly because there is such an institution as a university (characteristic for the Western culture) along with the technical (books as the material reproductions of knowledge) and economic infrastructure (a monthly scholarship).

The network determines to a large extent the scope of my actions, 'restricting' some of them while 'facilitating' others. Take the example of a simple direct interaction between a professor and students: an interactionist sociology perspective would analyze the situation using an interpretive framework. The interpretive framework defines the meaning and development of a given interaction.[6] In the case of an academic lecture, the framework is imposed by the institution of the university itself. We have been taught that within university facilities people play the roles of lecturers (those who should be listened to) and students (those who listen). ANT advances the idea of reconsidering the issue from a new perspective, paying special attention to the relations that enable and facilitate interactions. As may be guessed, a lecture is more complex than it might seem at the first glance (Latour, 2005, pp. 199-204). Firstly, the importance and outcome of the lecture hinges on a number of actions performed outside the lecture hall. Before the professor may begin to speak, the building needs to have a power supply, which is provided by a power plant. If for some reason the power is not being delivered, the lecture will have to be cancelled. Secondly, the network elements that are active during the lecture surpass its temporal location. The time passing in the course of regular social interactions is also a heterogeneous concept. It passes in one manner with regard to material objects (desks were produced five years before the lecture and possess characteristic durability), in another when it comes to the institution of the university (the lecture has to fit into the approved time schedule, otherwise it will be interrupted), and in yet another if we consider its very content (which may concern classical philology and works dated to times before Christ). Thirdly, not all the elements constituting the lecture are immediately visible. For instance, we do not see the power cables and are not aware of the components fitted inside the lecturer's computer. While creating a network of translations, these elements also exert influence on the progress of the lecture. Fourthly, the elements active during the lecture are not homogeneous. In fact, they may be material (desks at which the students sit), social (the aforementioned interpretive framework assigning social roles), and economic (financial standing of the university). Fifthly, not every element is active to the same extent. It may so happen that a microphone malfunctions during the lecture, and therefore it becomes the main actor shaping the outcome of the event. It is also possible that the weather gets worse and the rain pounding against the windows will make it impossible to continue the lecture. Thus, the weather may become an obstruction in accumulating knowledge.[7]

---

[6] "[...] in many cases the individual in our society is effective in his use of particular frameworks. The elements and processes he assumes in his reading of the activity often are ones that the activity itself manifests—and why not, since social life itself is often organized as something that individuals will be able to understand and deal with" (Goffman, 1974, p. 26).

[7] The lecture example has been used in order to make the reader aware that even a simple event is made up of complex elements. In the context of a scientist's life, a single

The network/context can also be analyzed from the global, instead of a local, perspective. It then turns out that the context, be it Ancient Greece or science of the fifteenth century, continues to be relational networks, not differing in that regard from the discussed interaction between the lecturer and students. This is because we always deal with objects, institutions, human beings, ideas, and practices of a well-defined and established nature, regardless of the extent to which these elements impact one another. In other words, the aim of ANT is to describe in a most detailed manner possible all the elementary particles of the context, which influence the behavior of the involved subject. Such complex entities as the Renaissance or the free market cannot be part of the explanation, but they need to be defined in relation to the studied biography. A history of the Renaissance can be written by analyzing academic culture. The culture, in turn, can be reconstructed with the help of histories of particular universities, which are linked with economic and political institutions, entailing another set of translations and actors. Finally, a history of the Renaissance can be based on the biography of Erasmus of Rotterdam, who attended a particular university in a particular country, met particular people, used particular objects, etc.

When dealing with a biography, the emerging translations resist any attempts to divide them into local and global ones (ibid., pp. 173-218). A biography is determined by a set of materials that extend beyond simple ontological, spatial, and temporal boundaries. In other words, human life is shaped by virtually all factors, which makes the task of choosing the most appropriate translations from their multitudes particularly challenging. In fact, it is one of the arguments cast against the Actor-Network Theory. As Olga Amsterdamska (1990) rightly observed, a researcher does not receive any tools which could be helpful in separating the significant networks from the non-significant ones. In the case of a biography, one ought to ask: is it necessary to analyze all networks in which the studied individual is involved? If yes, the researcher's task would become extremely difficult. In the end, it might even turn out that the biography resembles the Borgesian map, with the area amounting to that of the represented territory (Borges, 1972).

---

lecture does not have to prove decisive for his/her future. It does not have to—but it may. A good example here is Michel Foucault. It is commonly known that he delivered lectures at the prestigious Collège de France in the years 1971-1984. The lectures possessed certain characteristics: their total time had to amount to 26 hours, and the content was to be grounded in original research. As a result, the lectures have come to stand out as unique among the oeuvre of the author of *The Order of Things*, being a series of spontaneous ideas conceived in the lecture hall. What is interesting, the development and growing availability of tape recorders made it possible to publish the lectures as books. In that way, they have been functioning in the manner similar to the most important works by Foucalt (cf. Ewald & Fontana, 2003, pp. ix-xiv).

What can be done to solve this dilemma? A piece of helpful advice can be found in an article by a Finnish scholar Päivi Kaipainen (2010), whose suggestion is to begin by narrowing down the scope of the study. When analyzing a biography, we should ask ourselves, 'what is this process we are trying to capture?' Is it a network of dependencies leading to a particular scientific discovery? Or maybe it covers only the period before the studied individual became a well-known, outstanding researcher? Any restrictions we impose on ourselves at the very beginning will make it easier to navigate through the maze of heterogeneous translations. Another piece of advice, coming from Latour himself, is noticeably less optimistic. Scholars who describe networks always overlook some elements and the status of their work is prone to *rearticulation* (Latour, 2005, pp. 128-133). This means that by definition the analysis of a life story does not need to be an exhaustive study. A researcher creates a network of dependencies around the investigated individual with the help of the accumulated material. The level of the completeness of such description is more important than methodological accuracy, which in the case of ANT imposes the necessity to notice the relations between the elements. The focus of ANT lies predominantly with *action*. Thus, within a relational network, only those elements are meaningful that prompt actions of the investigated subject.

Network/Context: World War II, Hitler, Curie-Skłodowska, Uranium, Spies, Diplomats, and Twenty Six Containers of the Heavy Water—the Biography of Frédéric Joliot

To illustrate how the Actor-Network Theory pictures scientists' work, I will present the story of Frédéric Joliot, a French chemist, Nobel laureate, and husband of Irène Curie, the daughter of Marie Curie-Skłodowska. Joliot's input played a crucial role in the development of the atomic bomb by the Allies. His story is worthy of telling here for two reasons. Firstly, it will help us understand what it means that an individual operates within a context that cannot be grasped using only one general category, but rather dissolves into a number of details, dependencies, and active actors. In addition, it can also demonstrate why scrutinizing scientists' biographies is advisable when trying to explain the dynamics of science.[8]

---

[8] The following section tells a story reconstructed on the basis of three sources. The first one (Latour, 1999a, pp. 81-84) is representative of the ANT approach, and served as my main point of reference. The second one is a multi-plot story of the atomic bomb development, in which Joliot played one of many parts (Rhodes, 1986). Finally, the third source gives an account of the race between the Nazis and the Allies to get hold of the heavy water (Dahl, 1999, pp. 104-110); again, the French scientist was not the leading

As mentioned before, biographies provide resources necessary to debunk a few persistent myths regarding scientists' work.

Research on energy from uranium fission entered a critical stage in May 1939. With World War II looming large, Joliot managed to interest in his results both the French Ministry of War and the French National Center for Scientific Research (Centre National de la Recherche Scientifique—CNRS). Joliot desperately needed to forge contacts such as these as his experiments relied on uranium, whose supply was in deficit at the time. However, the issue was resolved with the help of the French government, since it turned out that Union Minière, a Belgian company, was extracting the element in a recently opened mine in the Congo. Union Minière mined uranium predominantly for the production of radium, which was much in demand in laboratories and medical centers all over the world after the discoveries made by Pierre Curie and his wife, Marie Curie-Skłodowska. Whereas the company did not have any use for uranium oxide, a by-product of the radium extraction, it was precisely what Joliot was after. According to the agreement negotiated by the French officials, Union Minière was obligated to deliver uranium oxide to France and to pay the scientist five million francs. In return, the Belgians were to receive 50% of the profits from all Joliot's patents.

Having secured the supply of the precious compound, Joliot and his two associates, Hans von Halban and Lew Kowarski set to work. Early on, they established that upon being bombarded with neutrons every uranium atom splits into two elements, releasing large amounts of energy. Joliot and his team tried to prove that the process of splitting could be turned into a chain reaction, with each atom produced as a result of splitting undergoing further bombardment with neutrons, generating further splitting and releasing even larger amounts of energy.

Even though the chain reaction was merely a hypothesis, it could be proven theoretically. Excited by the discovery, the French scientists were eager to share the news with the world. On the other side of the Atlantic, a Hungarian immigrant living in the United States by the name of Leo Szilard became concerned about their plan and tried to dissuade Joliot from publishing the article on chain reactions. Szilard had been deliberating

role. I have decided to extract a number of elements from different stories in order to illustrate the level of complexity and heterogeneity of the contexts in which Joliot was involved. I am fully aware that my choice has been arbitrary, and that sources always carry the risk of being problematic data. However, bearing in mind that the chain reaction story serves only as an example supporting my more general claim, I refrain from the in-depth analyses at this point. Otherwise, instead of discussing ANT in the context of researching scientists' lives, my paper would become an account-based analysis of Frédéric Joliot's story, which is virtually the same as writing a biography from scratch. Unfortunately, I have neither the skills nor the resources to attempt such a task.

over splitting atomic nuclei with neutrons since 1934, but had no idea as to which element would be best suited for the process. He was also familiar with Joliot's Nobel-winning work presenting the hypothesis that nuclear fission could produce vast amounts of energy. More importantly, the Hungarian scientist was convinced that discoveries in the field of nuclear physics would eventually lead to the creation of a deadly weapon. Burdened by the experience of his 1933 escape from the Third Reich, he also knew that the Germans would use the new tool to cause an unbelievable tragedy. To avoid this scenario, Szilard decided to patent his discovery in secret, communicating with the British and American governments. In addition, he also persuaded other scientists to conduct experiments on chain reactions in strict isolation from the exterior scientific world. The Hungarian had managed to enforce their code of silence until 22 April 1939, which was when Joliot, von Halban, and Kowarski published their article in the prestigious journal *Nature*. At that point everyone, including the Nazis, fascists, and Bolsheviks, intensified their efforts to build an atomic bomb. As the war grew close, ten parallel research teams were established in order to accomplish the objective.

After the outbreak of World War II, Joliot and his colleagues did not abandon the experiments. The main obstacle preventing them from producing an effective, sustainable, and safe chain reaction was excessive speed of neutrons bombarding the atom. To maintain fission, the scientists needed a suitable moderator that would slow down the neutrons. Perhaps the atomic bomb would have never been created but for the simple idea of Joliot's co-worker, Hans von Halban. His suggestion was to replace hydrogen found in water molecules with deuterium, which exhibited the same chemical behavior. The obtained compound became known as the heavy water. With the increase of the weight of water neutrons became heavier, their speed decreased, and the chain reaction was not interrupted. However, another issue arose. Owing to its low availability, obtaining deuterium, an isotope of hydrogen, proved extremely expensive.

As the war was raging on, French authorities began paying more and more attention to Joliot's work. This, however, did not win the sympathy of the left-leaning scientist. From the very onset of his career, Joliot was deeply convinced that his experiments should result in the production of a cheap energy source, not a nuclear weapon posing danger to people. Here I should mention the role of Raoul Dautry, an economist and engineer, who in the years 1939-1940 served as the French Minister of Armaments. His interest in the work on chain reactions had begun considerably earlier. The lives of the two gentlemen had intertwined at the end of 1939, before France was taken over by the army of the Third Reich.

The fact that the Nazis occupied Poland and intended to invade France as well as the difficulties in obtaining heavy water forced Joliot to reconsider his political views and reach a compromise with Dautry. While the scientist obligated himself to build an atomic bomb as soon as possible, the minister promised to provide him with a large supply of the necessary material. In Europe, the heavy water was produced exclusively in Norsk Hydro-Elektrisk in Norway. Upon Dautry's request, negotiations with the Norwegian company commenced. On that occasion, Deuxième Bureau, the French secret service, sent a spy by the name of Jacques Allier to Oslo. His task was to convince Norsk Hydro-Elektrisk to work together with the French. However, it was apparent that after Joliot's publication in *Nature* Germans also became intrigued by the heavy water and the Norwegian factory. Needless to say, Allier was acting in coordination not only with Dautry, but also with Joliot and the French president. He brought to the negotiations a check for 1.5 million kroner. The talks were successfully concluded on 9 March 1940. In the following days, two scheduled flights to France transported twenty-six specially made containers holding 185 kilograms of heavy water. Joliot, Halban, and Kowarski were able to continue their work, first in France, and later in England. The research and experiments conducted at that time led to the production of a chain reaction, construction of the bomb dropped on Hiroshima, and the post-war development of nuclear energy in France and all around the world. Owing to the scientist's exceptional skills, in 1945, general de Gaulle appointed Joliot the High Commissioner for Atomic Energy, and in 1948 he became the chief consultant for the construction of the first French nuclear power plant.

Let us now consider what measurable research benefits can be obtained from the Frédéric Joliot's story. To begin with, the story teaches us how to comprehend the action's context. As has already mentioned, ANT suggests viewing the relational network as a structure that determines the efficiency of its constituent elements. In this case, the acting subject of the analysis is Joliot, and the investigated process is the role that the French scientist played in developing the atomic bomb. However, contrary to the teleological assumption (stating that an autonomous individual seeks to achieve the goal set before commencing the action), the story of the chain reaction discovery seems to be a set of negotiations between a number of actors that exerted a measurable influence on the final result. Using the terms provided by the Actor-Network Theory, a series of translations had to occur in the challenging World War II conditions before the links critical to the final success could be created. The outbreak of the war itself proved to be crucial. It could be hypothesized that if it had not been for the rise of Adolf Hitler, chain reactions would have been used for peaceful purposes, as Joliot originally intended. To put it briefly, the outcome of Joliot's work could have been entirely different.

Let us consider for a moment what were the links that shaped Joliot's actions. First and foremost—a configuration of interests. Operating in a slightly different context, Michel Callon and John Law (1982) coined the term "map of interests" to denote those relations which are constituted by objectives of individual actors. The French government's interest was to build an atomic bomb and win the war. The Hungarian scientist Leo Szilard did a lot to prevent the publication in *Nature*. As any private company, Union Minière wanted to make money and fully exploit the potential of the Congo-based mine. Norwegian Norsk Hydro-Elektrisk focused their efforts on achieving profits by establishing the only facility in Europe capable of industrial-scale heavy water production. German spies sought to 'win' something for their superiors while their French counterparts tried to thwart their attempts. The relations among all these actions sketch out a map of interests across which Frédéric Joliot tried to navigate, aiming to trigger a sustainable and safe chain reaction. Some of the map elements had a direct influence on the scientist, such as when Raoul Dautry, acting on behalf of the French government, persuaded Joliot to act against his own political beliefs. Other elements gave rise to opportunities for achieving the goal, such as the case of Union Minière's unused uranium oxide, a by-product of radium production. Finally, there were also elements which Joliot simply could not ignore, such as Szilard's request. All these issues had an effect on the scientist's actions, and what is more, they were not the sole important factors in this context. In a sense, even Adolf Hitler, with his own set of interests that placed conquering the world at the forefront, influenced Joliot. It was the war unleashed by the Führer that forced the scientist to adjust his work to the French authorities' objective of defeating the Nazis.

Another type of links worth discussing concerns the economic, or, as ANT would call them, non-human factors. Their role in constituting networks was discussed on a number of occasions by the advocates of the Actor-Network Theory (Callon, 1986; Latour, 1992; 2005; Law, 1986a) and its commentators (Abriszewski, 2008b; Bińczyk, 2005). The most important argument put forward states that non-human factors, such as things,[9] standards, values, and elements of nature, are just as important as human

---

[9]   When considering these observations, it is worth noting that anthropologists inspired by, among other things, ANT advocate writing social histories of things, or even biographies of things (Domańska, 2008). Therefore, it is possible to imagine that the past of science could be analyzed not from the point of view of human actors, but foregrounding particular discoveries or inventions.

factors when it comes to triggering actions. Bearing in mind the biography and the relational context that a scientist has to deal with, what strikes me as particularly interesting are the negotiations with non-human factors in order to enable operations within particular relational networks (Latour, 1987, pp. 70-94; 1999a, pp. 174-215). Let us consider what actors Joliot faced. To become successful, in the privacy of his laboratory he had to negotiate the agency of neutrons by conducting a series of experiments. As mentioned above, their speed noted during the first split was too high, which resulted in the rupture of the chain reaction. The problem was solved only after an appropriate translation occurred, which here consisted in introducing a new actor (deuterium) and putting it in the place of the hydrogen molecule. Another good example are the containers in which the French spies, led by Jacques Allier, transported the heavy water. Since they were built in accordance with Joliot's detailed instructions, their special properties determined the successful outcome of the operation. While the material was being transported by scheduled flights heading to France, the containers turned into the leading actors that supported the success of the French scientist. As we may see, a non-human actor gives rise to certain problems (neutrons traveling too fast), which are solved by another non-human actor (heavy water) with the assistance of yet another non-human actor (special containers).

Let us reflect now on the benefits stemming from visualizing human life in terms of a relational network. In addition to capturing heterogeneous factors that influence the scientist, we are also offered a perspective that surpasses the individual. The notion of an individualistic nature of a scientific discovery, entailing that success is achieved by one particular scientist, is yet another myth deconstructed by the Actor-Network Theory.[10] Returning to Joliot's story, we may observe that his success was the outcome of cumulative efforts of his fellow scientists, Hans von Halban and Lew Kowarski, as well as his political allies: Raoul Dautry (representative of the French prime minister) and the spy, Jacques Allier. Joliot's colleagues assisted him in negotiations with neutrons, proposed a method of solving the problem of molecules' speed, and came up with an idea of translating the chain reaction experiment with deuterium as the key element. Relying on their diplomatic contacts, politicians forged a relation with the Belgian company Union Minière, and the designated spies showed their typical careful and tactful manner smuggling the containers

---

[10] Bruno Latour acknowledges that remarks on the collective nature of knowledge production can be found in the works of an eminent philosopher of science, Ludwik Fleck (Latour, 2005, pp. 112-114). As a matter of fact, the thought collectives theory (Fleck, 1979) is very similar to the Actor-Network Theory (Bińczyk, 2009).

with the heavy water without alerting the Germans. All of the agents acted, carried out translations, and maintained relations to allow Frédéric Joliot to successfully perform a continuous series of nuclear fissions and to be remembered in the history as the father of the French atomic energy.

Actor, or Acting in a Context, i.e. Networks

The relational network description presented above could be used to conclude that the context can subdue a person to such an extent that the individual cannot do virtually anything to control the course of his/her life. Nothing could be further from the truth. To portray a scientist's life following the Actor-Network Theory approach, it is essential to discuss the subject's mode of operation, translation, negotiation, and the way in which he/she influences other actors.

It is true that within the ANT framework the actor "is not the source of an action but the moving target of a vast array of entities swarming toward it" (Latour, 2005, p. 46), which could be seen in the case of Joliot. The actor can also be defined in a more subjective manner: "the actor makes changes in the set of elements and concepts habitually used to describe the social and the natural worlds. [...] [the actor] defines space and its organization, sizes and measures, values and standards, the bases and rules of the game—the very existence of the game itself" (Callon & Latour, 1981, p. 286). Being an active element, the actor (both human and non-human) can be so powerful as to constitute and 'organize' the surrounding environment in such a way that other actors will have to adjust to his/her rules.

The actor's actions can be described using the previously discussed concept of translation. Gathering actors within a network, modifying their attributes, and making them work for one's benefit are all included under translation, which can be understood as a practice determining the network's shape. Thus, establishing an actor's subjectivity consists first and foremost in weaving together numerous interconnected links. Let us refer once more to the academic lecture example. To be able to successfully deliver a presentation, the professor has to collect and control a number of heterogeneous elements, which include: mastering a given domain of knowledge (preparing the lecture for instance on the basis of his notes gathered through years of academic work); practicing specific speech techniques (so that the students do not doze off after 15 minutes of the lecture); and polishing a set of practical skills (writing on the blackboard, operating the microphone, communicating with the audience through computer presentations). The very fact of including the lecture into the curriculum is a result of the lecturer's efforts. He had to become employed by the university (fitting the university's demands for academic staff), push the idea for the lecture (getting the university administration interested in his achievements), and even negotiate the date of the lecture so that it would be acceptable for the speaker, the students, and other lecturers. Consequently, the actor/lecturer creates a network of translations. The lecturer's scope of work is translated into the will of employing a particular scholar at a certain university, while his interests are translated into the interests of students.

A relational network, as has been discussed earlier, is also prone to continuous rearticulation. The strength of its constituent links hinges of the techniques used for maintaining relations, attracting new actors, and adapting to the changing circumstances. Meanwhile, the potential of a given actor depends on his/her ability to transform the entire network. As network-constituting practices, translations do not so much attach new elements, but rather, by virtue of forming links change both themselves and the new elements. The basis of the Actor-Network Theory lies in continuous transformative actions. As Latour (2005, p. 45) observes, these actions often become unpredictable and unexpected: "Action should remain a surprise, a mediation, an event." Though seemingly insignificant, one actor can act in such a way as to establish global influence whose scope would far outreach the original, local context. The mobile phone can serve as a good example here. It may be viewed as a device typical of the information age of the Western world, allowing for the unrestricted transfer of messages, voice, and images regardless of the distance. At the same time, this small object has been transforming family relations, economic liaisons, emotional life, and work environment. Within each of these networks, it operates in an entirely different manner and is capable of yielding unexpected consequences. In contexts that are utterly different from the West in terms of culture, the device can be turned into a weapon, for instance by Islamic terrorists who detonate bombs using mobile phones.

Let us now consider some practical methodological suggestions concerning acting within relational networks (ibid., pp. 52-58). If the 'essence' of the actor are his/her actions, then this is the first aspect that should be discussed. According to the ANT guidelines, every action leaves a trace. The actor may write a diary, draft legal acts, write poems, or work on new inventions, all while keeping notes, having conversations, and so on. Each of these traces is at the same time the evidence of the individual's actions, as they illustrate how the given actor behaves in different contexts. A diary will produce a certain influence while being written, but will have a different effect after a few years' time. A similar observation applies to inventions: the printed word at the end of the fifteenth century behaved in a different way than it does now, in the Internet age. Following these traces is tantamount to following the actor, analyzing his/her actions and evaluating the degree to which he/she constitutes a network.

Secondly, special attention should be paid to what Latour calls "figuration" (ibid., p. 71). Every actor operates taking advantage of various forms and assuming different characters, which significantly expands the scope of his agency. In the discussed example, the mobile phone could serve as a communication tool in one situation, turn into a 'scaffolding' in a love game between two people, and become an effective weapon in another context. When considering the figure of a scientist, he/she may alternate between a positivistic ('I practice science for the truth') and a political one ('as a communist I will not focus on that'). Since each figure entails a different mode of acting, a great deal of thoroughness and carefulness is required to take into account the actor's mutable nature. Still, mutability with regard to the assumed figures is a desirable trait in actors. When operating within a heterogeneous network of dependencies, as demonstrated above, the actor has to freely navigate it. The liquidity of figures allows him/her to grasp the complexity of the world in which he/she functions, making him/her able to perform broader rearticulations, negotiations, and transformations.

Thirdly, actors that influence the network's shape, determine its actions, and define the agency of other actors continuously create antagonistic relations between one another. Actors fight among themselves, define counter-actors, deny the agency of others, and strive to accumulate such wealth of resources that would turn them into the dominating actor. A respected scientist who has achieved a high-ranking institutional position in a given field becomes surrounded by a network constituted so strongly that he/she is the one defining what it means to be a scientist. Thanks to the owned resources he/she can set the rules of the game.

It does not mean, however, that a strong actor plays the role of an absolutist hegemon, as every network can undergo some reconstitution. The process, known as the "trials of strength" (Latour, 1987, p. 78), denotes the moment when two actors meet in order to test the durability of each other's network. For instance, if two scientists compete to prove a given theorem, each of them participates in the duel by means of arguments, research instruments, and financial resources. It is a time when scientific controversies erupt, and only the final results will have the power to shape a given scientific theory. Importantly, the trial of strengths relies on translations that gather and acquire actors capable of defeating the opponent, i.e. creating a stable network that will withhold future attacks.[11]

---

[11] An emblematic example of trials of strengths is the conflict between Pasteur and Pochet, which Latour described in *Pandora's Hope* (Latour, 1999a, pp. 153-173).

The fourth factor that should be given special attention when studying an active actor is focusing closely on what the actors actually say. Latour (2005, p. 50) calls this stage "practical metaphysics." Every actor accepts different beliefs regarding the world in which he/she operates, as well as a particular model of behavior. It may be a realist ontology, and if a scientist were to embrace it, he/she would perceive his/her actions from the point of view of reaching the truth or capturing an entity in its realness. Conversely, a political ontology imposes a method of conceptualizing the world in accordance with a given ideology, i.e. a set of ideas concerning collective life. What is equally important is the actor's self-definition as an acting element. A scientist convinced of the significance of his/her work will behave in one way (defining his/her subjectivity as exerting real influence on the surrounding world), while an amateur inventor locked in his/her workshop will be quite different in this regard (being devoid of ambitions to become a scientific revolutionist).

Let us return to life stories of scientists. As has been indicated before, a relational network exerts influence over an individual by means of translation, which in turn causes behavioral changes. Some elements help the subject achieve its targets, while other function as obstacles. My intention, however, was neither to deprive the scientist of his/her subjectivity nor to knock him/her off the pedestal to which he/she was raised by the history. In the context of the Actor-Network Theory it is possible to distinguish outstanding individuals, remarkably creative ones, and geniuses. Rather than being determined by an isolated entity with a solitary will, these characteristics are governed by a series of mobilized relations. In other words, an outstanding individual, firstly, is capable of creating a large number of links; secondly, he/she is able to skillfully maintain them; and thirdly, he/she has the capacity for exerting a considerable influence on other actors and their relations (the greater the impact, the more outstanding the individual) (see Chapter Five in this book).

Building relations, as I have already stated, stands for weaving networks. This involves gathering allies and spokespersons, acquiring nonhuman factors, identifying and rearticulating interests in a particular context. On the other hand, maintaining relations covers all the practices aimed at moving the actor to a dominant position—and they also are contingent on successful translations. Finally, exerting an influence entails having an impact, setting the rules of the game, as well as modifying and rearticulating the existing networks. It is worth noting that this influence escapes the restrictions that are not only spatial, but also, more interestingly, temporal in nature. According to Latour (1993, p. 74), "time is not a general framework, but a provisional result of the connection among entities." The time that a scientist has for acting does not need to be seen as a closed period between his/her birth and death. Rather, it is a set of relations between a given discovery and all the relational systems into which the discovery will later be embedded.

Actor in a Network/Context: a Double Helix, Linus Pauling, Cardboard Cut-Outs, Mysterious Data General Project, and a 32-Bit Computer, or Biographies of Francis Crick, James Watson, and Tom West

To illustrate the way in which an actor/scientist operates in a given context/network, I will once more give a rather elaborate example. This time it covers the stories behind the invention of the DNA double helix model and the Eclipse MV/8000 minicomputer.[12] Although quite different at the first glance, these accounts have much in common; in fact the stories come together in the late 1980s, when advances in computer science contributed to unprecedented discoveries in the research on the human genetic code.

Let us go back to 1951. In the Cavendish Laboratory in Cambridge, two scientists, Francis Crick and James Watson, are working on the first model of the deoxyribonucleic acid in the history of science. At that stage, it is still not known whether they are dealing with a double or a triple helix, and whether the phosphate bonds are inside or outside the molecule. Seeking to solve the puzzle, Watson and Crick follow the paths of contemporary science, trying to obtain the DNA structure using X-rays. Meanwhile, a renowned chemist and physicist residing in the United States, Linus Pauling, announces that he is close to revealing the DNA structure, and that the project should be completed within a few months. To make things worse, Sir Lawrence Bragg, the supervisor of the English scientists, does not share their enthusiasm and advises them to focus on more serious matters.

Pauling makes a discovery. He postulates that the DNA structure is a triple helix with a sugar-phosphate backbone in the centre. A scientist and friend of Watson and Crick brings them the American's article before the publication. At first, both are furious at their superior. Had he not

---

[12] I reconstruct portions of the rather complicated stories relying predominantly on Latour's *Science in Action* (1987, pp. 1-13), where he uses them to illustrate his own claims. His work is based on Tracy Kidder's *The Soul of the New Machine*, a detailed chronicle narrating the struggle of Tom West and his colleagues with the prototype Eagle under the conditions offered by his company, Data General. For the DNA story, the main source was a book by James Watson himself, entitled *The Double Helix: Personal Account of the Discovery of the Structure of DNA*, to which I also refer (Watson, 1968).

stopped the research, maybe they would have achieved their goal sooner than Pauling. However, upon closer inspection of the triple model, the scientists make a surprising discovery. Pauling did not include hydrogen atoms in any of the three chains, which defied the fundamental laws of chemistry—a model without hydrogen would not hold together. Watson and Crick realize that the American committed a schoolboy error, which means they can still make a revolutionary discovery. What they also realize is that the moment the article is published in the prestigious *Proceedings of the National Academy*, the mistake will be immediately detected, and Pauling will continue his work on the DNA structure. Watson and Crick know the article will appear in six weeks, which means they need to hurry.

Encouraged by the failure of his American colleague, Watson ponders upon the DNA structure, taking into account its many variants. In popular chemistry textbooks, he encounters a principle defining tautomeric forms and notices a surprising symmetry in the structure of nucleic acid: adenine corresponds to adenine, cytosine to cytosine, guanine to guanine, etc. However, Watson does not know that the tautomeric forms he has found are wrong. He probably would have never learned that if it had not been for the fact that in those times he shared his office with Jerry Donohue, an American chemist who came to Cambridge on a six-month grant from the Guggenheim Foundation. Donohue tells Watson that the model relying on tautomeric forms, found in the classical James N. Davidson's textbook, was not based on a reliable research. As a replacement, the American suggests using the model applying keto forms and obtained through more thorough analyses.

Watson does not have to take his advice. The American is only a visiting scientist from the outside world and not a member of the research collective. Besides, he used to study with Pauling, the main rival in the race to discover the DNA structure. Still, Watson decides to follow his advice. He locks himself in the laboratory, makes cardboard cut-outs of the elements found in the model of the deoxyribonucleic acid, and tries fitting them together. Thanks to the visual presentation and the use of manual skills, the British scientist succeeds in building a working model. Watson makes sure that it complies with the fundamental laws of chemistry. Donohue and Crick confirm that the model is correct. As it turns out, the cardboard cut-out double helix represents the actual DNA structure.

Nearly 30 years later, in the Data General facilities in Massachusetts, we find Tom West and his team trying to eliminate design flaws in the Eagle minicomputer prototype. Data General has great hopes for the invention because DEC, a rival company, has recently began selling their VAX 11/780 model. Work on the new computer has been delayed due to the failure on the part of the manufacturer (supplying special PAL processors that were to be fitted in Eagle) to ensure a timely delivery. Just like Watson and Crick, West is not supported by his superior, de Castro, who being disappointed by the recent unsuccessful project Ego (also managed by West) has been considering providing support to an entirely new enterprise in North Carolina, where a rival team operates.

Considering the unfavorable circumstances, West probably would not have been as motivated as he was if it had not been for a peculiar event. One day, a colleague working for DEC takes him in secret to the company's basement to show him the VAX 11/780 model. At that point West realizes that his competitors have assembled a working, but highly inefficient and expensive computer. West knows the organizational structure of DEC and is aware of the fact that it is a rather bureaucratic, conservative company that does not take unnecessary risks. This policy is clearly reflected in the design of their flagship product, VAX 11/780.

West decides to take the risk. Not intimidated by the delay in comparison to DEC, the newly-established rival group in North Carolina, and the failure of the previous project, West believes that he will be able to design a computer that is more efficient, faster, and cheaper. To do so, he isolates his team from the rest of the company, making sure that his colleagues would be able to work on the revolutionary invention without being bothered. In short, he creates a new collective and hides it within the structure of Data General. The task is far from being easy. West has to supply his immediate subordinates with appropriate materials and funding while conducting negotiations and acting as the team representative in front of his superior, the North Carolina group, and the marketing department that is predominantly focused on gaining quick profits.

The enterprise lasts two years. For 24 months West keeps the project in secret, pretending that his team is busy with an altogether different task. In addition, he lobbies his superior and the marketing department, in that way obtaining the resources necessary for achieving the goal. With time, West engages in the project the most crucial sectors of the company. As the North Carolina team suffers a failure (unable to design a new computer), and with DEC becoming ever more powerful and competitive, project Eagle becomes the last hope of Data General. As a result, expectations towards the members of the West's team are growing. Inevitably, there comes the moment when the clandestinely working team will have to demonstrate the final product. However, new issues arise continuously: Eagle operates in a stable way only for a few seconds, and the PAL processor manufacturer is on the verge of bankruptcy. West has to agree to let experts in software and hardware diagnostics into his laboratory. The efforts of the whole team of computer design specialists, who for weeks have continuously numerous flaws, finally result in the development of the 32-bit Eclipse MV/8000 microcomputer.

These stories come to a climax in 1985. In that year, John Whittaker, working at the Institut Pasteur in Paris, develops a computer program for analyzing DNA. To do so, he takes advantage of the computer assembled by West and his team, as well as the double helix model proposed by Watson and Crick. At this point, the two inventions triggered an avalanche of discoveries linked with the human genetic code. The 1980s witness a surge in the DNA sequencing research, which involves determining the order of the nucleotide pairs. Thanks to the combination of the computer developed by Data General and the discovery of the double helix, it has become feasible to describe the entire human genome, which gives rise to inconceivable cognitive benefits.

What conclusions could be drawn this time about the scientist's activity? First of all, the story presented above reveals active agency. Although the network/context demarcates the scope of activity (as has been mentioned before), it is the decisions taken by the individuals that have the greatest significance. Watson did not have to listen to Donohue simply because he might have mistrusted him. West could have refused to let external specialists enter his laboratory and attempted to finish the project on his own. The moments when these seemingly insignificant decisions were made became decisive to the given scientist's success when considered in the context of the whole story.

Second the outcome achieved by both West and the DNA researchers was to a large extent dependent on skilful gathering of resources. To reach the goal, a successful chain of translations needs to be created in such a way so as to make heterogeneous elements operate as one entity. What was it that our individuals gathered? To begin with, all the non-human elements associated with their work. Watson and Crick had to combine nucleic acid components observing the laws of chemistry and introduce hydrogen into them, the sugar-phosphate backbone, and helices. On the other hand, West had to solve issues stemming from the use of faulty software and hardware.

This still does not exhaust the list. Watson, Crick, and West had to sort out a number of arrangements with their superiors, institutions, and companies. Engaging oneself in science- or technology-related activity does not involve solely designing experiments and inventions within the safe confines of closed research institutes, but also covers generating interest about the project, and acquiring financial resources and equipment. This is best illustrated by the example of West. In order to establish a comfortable work environment for his team, he had to become the collective's representative. This obligated him to review their progress in front of the marketing department, software department, and, more importantly, in front of his superior, de Castro. But for West's diplomatic skills, Eclipse MV/8000 would probably have never been built.

Acquiring funds for activity is linked with an issue known in ANT as *translation of interests*. As has been signaled earlier, the context/network is co-created, among other things, by a map of interests, which stands for the collection of different actors' objectives. To perform a successful translation, and also achieve one's goal, the interests should be identified and rewritten in such a way so as to make all the involved parties aware that they are acting toward their personal goal. As Latour (1999a, p. 88) notes, "translations consist of combining two hitherto different interests [...] to form a single goal." West sought to build Eclipse MV/8000; the Data General marketing department wanted to offer their customers the best possible product; the boss did everything in his power not to incur a loss of profits; the dedicated, young, but at the same time inexperienced computer science specialists had a unique chance to demonstrate their skills. To combine these largely dispersed goals into one objective and subsequently use it to advance the project, a series of negotiations, transformations, and gathering the spokespersons had to be carried out. As we have seen, West coped brilliantly with the task. He managed to convince the marketing department that his computer would be cheaper and more efficient that the one offered by DEC. Choosing a suitable moment (after the failure of the North Carolina team), he also persuaded his superior that project Eagle was the company's last hope. Finally, he let into his laboratory a group of promising specialists, who worked day and night on removing design flaws.

When studying biographies, it is vital not only to pay attention to the acts of mobilization and translation, but also to notice what the actors themselves have to say about their actions. The third element that can be observed in the stories of Watson, Crick, and West is the way in which their conceptualization of both the actions taken up by them and the environment influenced their behavior. We would not have been able to comprehend Watson's decision to follow Donohue's advice if we had not known he considered the American to be an expert in the field. Similarly, we would not have understood West's decision to take the risk of working for two years in secret if we had not been aware of the fact that he knew about the flaws of the computer designed by the rival company.

The fourth point which should not be overlooked is the creation of a dense relational network that expands well beyond the contexts of the DNA discovery or constructing Eclipse MV/8000. Let us not forget that one of the indications of how outstanding an individual is lies in the number of relations he/she has managed to generate around himself/herself, thus exerting an influence capable of impacting other actors. A powerful actor is an actor who defines the rules of the game, establishes the framework for actions, and determines the standards as well as areas of objectivity. To use a term occasionally employed by ANT (Latour, 1987, p. 132; Law, 1986b, p. 34), an outstanding individual sets "obligatory points of passage." Once the double helix has been discovered, every scientist researching DNA has to study the structure proposed by Watson and Crick, and conduct his/her research in accordance with their guidelines. Therefore, the double helix has become an obligatory point of passage which needs to be crossed before making further progress.

Still, there is something more to add. The discoveries of the British scientists as well as the American invention influence networks which seem to be far from their original contexts. The DNA helix is present not only within the fields of molecular biology and genetics, but also in medicine and forensics. Hence, the agency of Watson and Crick makes an impact touching not a narrow group of specialists, but also people dealing with healthcare or prosecuting offenders. Doctors and detectives alike have to cross the obligatory point of passage established in 1953. The networks of influence could be traced even further. By accepting the double helix discovery as a scientifically proven fact, each of us conceptualizes and visualizes deoxyribonucleic acid following the model of Watson and Crick. When thinking about DNA, the first image coming to mind is that of the two interwoven ribbons.

Conclusions, or Why it is Worth Following the Actors

Although I have chosen separate stories to illustrate two issues (*the context* and *acting within the context*), they are not markedly different from the Actor-Network Theory perspective.[13] If there is no distinction between the actor and the network, the acting subject remains embedded in the context of his actions. While the actor is the network and the subject is the context, the scientist creates a series of relations, determines the scope of other actors' actions, sets the rules of the game, and establishes obligatory points of passage. Owing to all these practices, rather than being an individual scientist, he/she becomes a set of non-human factors, practices, financial resources, institutions, etc. At the same time, the network is the actor, and

---

[13] "'Actor' does not play here the role of subjectivity, and 'network' does not play the role of the society. Actor and network [...] mean two sides of the same phenomenon [...]" (Latour, 1999b, pp. 18-19).

the context is the subject—networks of dependencies continue to function, triggering modifications and negotiations, creating frameworks for actions, multiplying some possibilities while doing away with others.

Let us conclude. The life story of a scientist is far more complex than has been traditionally viewed by the philosophy of science or the classical sociology of knowledge. As suggested in the introduction, scientific activity is subject to mythologization, which stems from the exceptional status granted to science in our culture. As a consequence, a scientist is perceived as a lone genius, a creative individuality who can reach his/her intended goal. Interpreting the role of the scientist along these lines obscures the real picture of how knowledge is generated. If we examine all the intricate elements (cultural, social, financial, political) involved in conducting scientific work, it will be possible to adopt an altogether different approach. Another analysis of the stories presented above can make us aware of the fact that no absolute divisions exist between *the content of science* and its *context*. The division postulates that a scientist, possessing innate rationality, operates within the former and merely exists in the latter, without exerting any measurable influence on his/her cognitive activity. As I have attempted to show, such division is artificial. It is not possible to separate interests, values, ideas, things, and politics from neutrons, heavy water, uranium, radium, guanine, cytosine, and sugar-phosphate backbone.

This is precisely the lesson to be learned from ANT and transferred into the field of traditional reconstruction of scientific knowledge: no absolute separation exists between factors that are social and natural, material and non-material, human and non-human, political and non-political, subjective and objective. Originating in the dynamics of science, that is in a series of transformations, negotiations, and rearticulations, a dynamic relational network emerges, defying simple ontological distinctions. However, the Actor-Network Theory would not prove particularly useful for the history of science if it were not for the previously collected examples. Thanks to well-written biographies, the heterogeneity of scientists' actions reveals its true extent before our eyes. Obviously, ANT cannot evaluate biographies in terms of being better or worse from a historical point of view. There are no tools suitable for determining which biography uses more adequate sources and represents the past in a more accurate manner (this type of analysis belongs to the domain of historical research, which possesses tools fitting the purpose). Nevertheless, ANT makes it possible to evaluate their usefulness in explaining the complexities of research processes. To do so, it employs a number of biographies which facilitate the creation of a relevant network. In the case of the atomic bomb example, one network would explain Joliot's success, another the role of Szilard, and yet another could analyze Einstein's influence. It is also possible to rewrite the story in such a way as to place the atomic bomb itself in the centre.[14] Each of the networks can use different sections of the biography.

Whenever analyzing a biography with the aim of evaluating the status of a scientific discovery, we should observe the basic methodological advice provided by ANT: follow the actors (!) (Latour, 2005, p. 68). This is the heuristic profit that can be obtained by means of the Actor-Network Theory. Studying life stories and biographies of scientists in order to find the explanation of the dynamics of science in terms of relational networks, makes it possible to observe the practice of translations, which in turn contributes to a deeper understanding of the actors' actions. We can dig out the origin of a given discovery, and also understand what the motivation was behind the agency of the involved actors and what the consequences were of such discovery. By adopting this approach, we reject the naive belief in the superior role played by an entirely rational, individually acting subject in the process of knowing. According to the ANT model, the scientist continues to work and to be an outstanding individual. The difference consists in that his/her exceptionality is no longer defined by a solitary will following a straight path to a revolutionary discovery, but rather by a set of relations, allies, and practices that the actor is capable of mobilizing.

 C8

Acknowledgments: I would like to thank Justyna Zielińska for her help in writing this article. Without her valuable comments, it would have been an entirely different (worse) text.

References

Abriszewski, K. (2010). Splatając na nowo ANT. Wstęp do *Splatając na nowo to, co społeczne*. In B. Latour, *Splatając na nowo to, co społeczne* (pp. v-xxxvi). (A. Derra & K. Abriszewski, Trans.). Kraków: Towarzystwo Autorów i Wydawców Prac Naukowych Universitas.

Abriszewski, K. (2008). *Poznanie, zbiorowość, polityka: Analiza teorii aktora-sieci Bruno Latoura*. Kraków: Towarzystwo Autorów i Wydawców Prac Naukowych Universitas.

Abriszewski, K. (2008b). Rzeczy w kontekście Teorii Aktora-Sieci. In J. Kowalewski, W. Piasek, & M. Śliwa (Eds.), *Rzeczy i ludzie: Humanistyka wobec materialności*

---

[14] This approach has been adopted by the previously mentioned Richard Rhodes (1986).

(pp. 103-130). Olsztyn: Series: Colloquia Humaniorum. Olsztyn: Instytut Filozofii Uniwersytetu Warmińsko-Mazurskiego.

Amsterdamska, O. (1990). Surely, You Are Joking, Monsieur Latour! *Science, Technology, & Human Values, 15* (4), 495-504.

Bińczyk, E. (2009). *Praktyka, laboratorium, czynniki pozaludzkie: Najnowsze modele technonauki oraz wybrane tezy Ludwika Flecka,* http://fleck.umcp.lublin.pl/teksty.binczyk2009.htm [last accessed: March 9, 2011].

Bińczyk, E. (2007). *Obraz, który nas zniewala: Współczesne ujęcia języka wobec esencjalizmu i problemu referencji.* Kraków: Towarzystwo Autorów i Wydawców Prac Naukowych Universitas.

Bińczyk, E. (2005). Antyesencjalizm i relacjonizm w programie badawczym Bruno Latoura. *Er(r)go, 1,* 91-101.

Borges, J.L. (1972). Of Exactitude in Science. In J.L. Borges, *A Universal History of Infamy* (p. 141). (N.T. di Giovanni, Trans.). New York: E.P. Dutton & Co.

Callon, M., & Latour, B. (1981). Unscrewing the Big Lewiathan: How Actors Macrostructure Reality and Sociologists Help Them to Do So. In K. Knorr-Cetina & A. Cicourel (Eds.), *Advances in Social Theory and Methodology: Toward an Integration of Micro- and Macro-Sociologies* (pp. 277-303). London: Routledge & Kegan Paul.

Callon, M. (1986). Some Elements of a Sociology of Translation: Domestication of the Scallops and the Fishermen of St Brieuc Bay. In J. Law (Ed.), *Power, Action and Belief: A New Sociology of Knowledge?* (pp. 196-229). London: Routledge & Kegan Paul.

Callon, M., & Law, J. (1982). On Interest and Their Transformation. *Social Studies of Science, 12,* 615-625.

Czerniak, P., & Węgrzecki, A. (1990). Wstęp. In M. Scheler, *Problemy socjologii wiedzy* (pp. VII-XXV). Warszawa: Wydawnictwo Naukowe PWN.

Dahl, P.F. (1999). *Heavy Water and the Wartime Race for Nuclear Energy,* Bristol – Philadelphia: Institute of Physics Publishing.

Domańska, E. (2008). Problem rzeczy we współczesnej archeologii. In J. Kowalewski, W. Piasek, & M. Śliwa (Eds.), *Rzeczy i ludzie: Humanistyka wobec materialności* (pp. 27-60). Series: Colloquia Humaniorum. Olsztyn: Instytut Filozofii Uniwersytetu Warmińsko-Mazurskiego.

Ewald, F., & Fontana, A. (2003). Foreword. In M. Foucault, *"Society Must Be Defended": Lectures at the Collège de France, 1975-1976* (pp. ix-xiv). (D. Macey, Trans.). New York: Picador.

Fleck, L. (1979). *Genesis and Development of a Scientific Fact*. (T.J. Trenn & R.K. Merton, Eds.; F. Bradley & T.J. Trenn, Trans.). Chicago – London: University of Chicago Press.

Goffman, E. (1974). *Frame Analysis: An Essay on the Organization of Experience*. New York: Harper & Row.

Havelock, E.A. (1963). *Preface to Plato*. Cambridge, MA: Basil Blackwell.

Kaipainen, P. (2010). *Actor-Network Theory in Biographical Analysis*, http://www.allacademic.com//meta/p_mla_apa_research_citation/1/7/7/3/4/pages 177343/p177343-1.php [last accessed: March 9, 2011].

Kuhn, T.S. (1957). *The Copernican Revolution*. Cambridge, MA: Harvard University Press.

Latour, B. (2010). *Splatając na nowo to, co społeczne*. (A. Derra & K. Abriszewski, Trans.) Kraków: Towarzystwo Autorów i Wydawców Prac Naukowych Universitas.

Latour, B. (2005). *Reassembling the Social: An Introduction to Action-Network Theory*. Oxford – New York: Oxford University Press.

Latour, B. (1999a). *Pandora's Hope: Essays on the Reality of Science Studies*. Cambridge, MA: Harvard University Press.

Latour, B. (1999b). On Recalling ANT. J. Law & J. Hassard (Eds.), *Actor Network Theory and After* (pp. 15-25). Malden, MA: Basil Blackwell.

Latour, B. (1993). *We Have Never Been Modern*. (C. Porter, Trans.). Cambridge, MA: Harvard University Press.

Latour, B. (1992). Where Are the Missing Masses? The Sociology of a Few Mundane Artifacts. In W.E. Bijker & J. Law (Eds.), *Shaping Technology/Building Society: Studies in Sociotechnical Change* (pp. 225-258). Cambridge, MA: Massachusetts Institute of Technology Press.

Latour, B. (1987). *Science in Action: How to Follow Scientists and Engineers through Society*. Cambridge, MA: Harvard University Press.

Law, J. (1992). *Notes on the Theory of the Actor Network: Ordering, Strategy and Heterogeneity*, *www.lancp.ac.uk/fass/sociology/papers/law-notes-on-ant.pdf* [last accessed: February 21, 2011].

Law, J. (1986a). On the Methods of Long Distance Control: Vessels, Navigation, and the Portuguese Route to India. In J. Law (Ed.), *Power, Action and Belief: A New sociology of Knowledge?* (pp. 234-263). London: Routledge & Kegan Paul.

Law, J. (1986b). On Power and Its Tactics: A View from the Sociology of Science. *The Sociological Review, 34*, 1-38.

Rhodes, R. (1986). *The Making of Atomic Bomb.* New York: Simon & Schuster.

Sojak, R. (2004). *Paradoks antropologiczny: Socjologia wiedzy jako perspektywa ogólnej teorii społeczeństwa.* Wrocław: Wydawnictwo Uniwersytetu Wrocławskiego.

Watson, J.D. (1968). *The Double Helix: A Personal Account of the Discovery of the Structure of DNA.* London: Weidenfeld & Nicolson.

Zybertowicz, A. (1995). *Przemoc i poznanie.* Toruń: Uniwersytet Mikołaja Kopernika.

Chapter Four

THINKING—COMMITMENT—ASSIMILATION

BIOGRAPHICAL THEMES IN THE SOCIOLOGY OF ZYGMUNT BAUMAN

by Magdalena Matysek-Imielińska

The popularity and power of social persuasion of Zygmunt Bauman are due mainly to books written in the 1980s and early 1990s. *Legislators and Interpreters*, *Modernity and the Holocaust*, *Modernity and Ambivalence* and *Postmodern Ethics* are all publications which put Bauman in the midst of the prolific and diverse discussion about postmodernity. It was the period when the most heated debate on the state of the modern world was taking place (or, in fact, drawing to a close—since it was rather the time of conclusions). Bauman's readers saw in his books a critique of postmodernity, a new vision of the coming times, frequent warnings concerning the freedom of choice, pluralism, but also making us realize that postmodernity—with its diversity, lack of faith in the reason, science and great social projects—gives us a chance that can be used by anyone who is able to abandon the illusion of modernity.

Everyone read Bauman similarly then, thinking that he just speaks on a matter that is currently valid. However, if we were willing to do the reading of these books in terms of a more historical or biographical perspective, we would need to pose a question about the personal experiences of the author, allowing him to make such a reflection. Adopting such viewpoint is all the more legitimate since today, more and more often, references are put forward to Bauman's biography or the autobiographical writing of his wife, Janina; what is more, he himself, talking about his own life is eager to put it in the context of the history of modernity. Also significant in this respect is, for instance, Bauman's view on the parallel nature of his own way of life with the lots of Cornelius Castoriadis. This similarity refers not just to a convergence of views and political ideas, but above all to the weaving of individual experience into historical experience, a certain idea of a generation; it's about "the similarity of that curious and difficult to disentangle mixture of continuity and discontinuity" (Bauman & Tester, 2001, p. 36). Bauman's biography reminds an intellectual chronicle of an era, where the biography is intertwined with the history (of the country, generation, ideas, utopias). What Bauman writes about Castoriadis ("At no stage did he attempt to steer clear from the concerns, worries and ambitions of his contemporaries: never did he seek to locate his own interests at a safe distance from the main intellectual battlegrounds of the time" (ibid., p. 37)) can be successfully applied to him: throughout all his long life he has never stayed aside (emigration to the Soviet Union in 1939, joining the Polish First Army in 1943, membership in the Polish Workers' Party and then PZPR[1]) flirting with Marxism, Marxist revisionism and humanist socialism, practicing engaged sociology or taking up criticism of modernity and capitalism.

Bauman's works are "like an intellectual chronicle of the age, a faithful record of successive generations' experiences, discoveries and blind spots, hopes and disappointments, naivetés and wisdoms" (ibid.), so it is impossible to receive them without reference to the thread of individual experiences. Undoubtedly, the vision of this kind of sociology is the result of a deliberate life strategy. In this sense, Dennis Smith (1999, p. 3) says, "Bauman is part of the story he tells. His can be found on the map he draws." This is evident, for example, when Bauman marks out the tasks of sociology: "we would like [sociologists—M.M.-I.] to show us how our *individual* biographies intertwine with the *history* we share with fellow human beings" (Bauman, 1990, p. 10; italics in the original). The demand of combining biographical themes with broader social thought is tempting, and also justified, especially when, in relation to the biography that is of the highest interest to us here, we consider the configuration of the following circumstances:

1. Zygmunt Bauman, over time, has been torn in his views between faith (especially important in this regard seems to be the Polish period

---

[1]   Polska Zjednoczona Partia Robotnicza—Polish United Workers' Party was formed following the merging of the Polish Socialist Party (PPS) and the Polish Workers' Party (PPR). The union of these parties meant that since then there existed in Poland a one-party system. Below I will mostly use the name 'the Party,' except where original form needs to be preserved in citations.

of his Marxist intellectual activity) and doubt that—lined with personal experiences—related to the systemic nature of the social order as a modern illusion; the new hope to get close to the socialist utopia is brought about by emigration that gives Bauman the opportunity to observe from the outside the effects of the emancipatory activity on the part of workers, creation of the trade unions, forming the resistance, etc.

2. An important part of Bauman's discourse related to the criticism of modernity occurs on the level of re-working through the family experiences of Janina (cf. J. Bauman, 1986). *Modernity and the Holocaust* comprises an analysis of modernity as a system of rationality and bureaucracy, in which opportunities and trends of modernity culminated; an attempt of handling the things that go beyond the common thought, and that—despite the fact that the Holocaust is already a part of the past for them— could happen again; this work reveals a dilemma of a human being trying to understand their own position in the face of a struggle with the new, inhumane system.

3. Yet another aspect of biographical references is related to the ambivalence as a condition accompanying Jewish intellectuals. *Modernity and Ambivalence* highlights the treachery of the pitfall of assimilation (Bauman fell into it as a Jew living in the post-war Poland) and the trap of commitment (relating to the intellectual realm); when a diagnosis is made of the latter it seems helpful to read *Legislators and Interpreters*.

4. It is also worth considering (without expecting simple conclusions) whether *Postmodern Ethics* can be regarded as a kind of a warning for future generations, fascinated by the prospect of social paradise and moral happiness; a warning against ethical principles that are too easy and give a sense of security precisely defining the boundaries of good and evil, friends and enemies. Such warning was uttered by Janina Bauman in her memoir concerning the years after the war titled *A Dream of Belonging: My Years in Postwar Poland*[2] (1988; this book was also published in an altered Polish version under the title *Nowhere on Earth: Returns—Stories* (*Nigdzie na ziemi: Powroty—Opowiadania* (J. Bauman, 2011)).

Would Zygmunt Bauman, who took part in the then division of the world into the enemies and friends of the system, be able to get rid of his own experiences from the scientific discourse? And maybe—after all—it will be easier for us to understand the message contained in Bauman's texts when we connect reading them with the biography of the author?

---

[2]   We read there: "If you want to do great things which you cannot do on your own, remember that NOBODY will ever exonerate you from the responsibility of an unexpected outcome of the collective action. This is the message I leave to my children and grandchildren as my last will and testament" (J. Bauman, 2011, p. 72; form as in the original).

Keith Tester, among others, is wary of such way of reading Bauman's works as in his opinion, this way involves serious reductionism, which confines social thought only to an individual dimension. Tester would also like to avoid—for the reasons indicated by Richard Sennett in *The Fall of Public Man*—veiling *the public sphere* with *what is private* (cf. Bauman & Tester, 2001). It seems that concerns of this type are relatively easy to resolve, if we take a biography as a chronicle of an era and combine it in this case with the ideas of modernity, the realities of the pre-war and the post-war Poland, the historical background of Stalinism etc. Valuable tips keep flowing to us from Charles Wright Mills, who in *The Sociological Imagination* calls for placing the fate of individuals in a wider social context in order for the fears, anxieties and concerns of a particular human being to be raised to the category of a public problem, so that they are recognized in terms of major issues for the whole generation. Turning individual anxiety into collective commitment is one of the most important challenges of sociology; sociological imagination that allows people to understand what is happening or what has happened outside their nearest environment also provides the opportunity to gain self-awareness that builds us.

In the 'Trap of Commitment'

Maria Hirszowicz, whose biography has some similarity to that of Bauman (she was e.g. an active member of the Party and a lecturer at the Warsaw University, she published texts of praise about Marxism and, ultimately, she made a strong revision of her earlier views, to finally be subjected to forced emigration),[3] asks herself how it was possible that soon after the war, with the Nazi totalitarianism still fresh in their memory, Polish intellectuals relatively easy rushed to embrace the 'new faith'? Hirszowicz tries to answer this question in a semi-autobiographical book, *Pułapki zaangażowania: Intelektualiści w służbie komunizmu* (*Traps of Commitment: The Intellectuals in the Service of Communism*). The author notes that the faith in the communist system was built based on an ideal vision of the society of equality, freedom and solidarity. The undeniable charm of Marx's theory probably resulted from a potential concealed in it for a critical look at the status quo, which in the case of the Jews that had survived the war had special significance. Striving to overcome social isolation, quite visible

---

[3]   In 1968, the Polish communist government of Władysław Gomułka carried out a propaganda operation against Polish citizens of Jewish origin. Gomułka removed then Polish-Jewish professors associated with revisionism from universities. These were Leszek Kołakowski, Bronisław Baczko, Zygmunt Bauman, Maria Hirszowicz and others.

here, made itself more real through the existence in the community, bearing signs of mutual loyalty and emotional support. From this, there was only a step to automystification (self-deception), emerging at the verge of ideo-logical commitment and truth. "Can you take a stand on important social issues and play an active role in the public life without falling into the trap of automystification"?—wonders Hirszowicz (2001, p. 8). The truth always depends on the circumstances, it facilitates the control of the situation and brings people closer to the purpose. In the important social issues, the truth is not achieved through a procedure that leads to the purification of the 'accretions' of external circumstances. Such truth can be sought on scientific grounds, but not in matters of human slavery and experienced harm. Then, the truth is easily changed into an ideology, and the conflict between them leads inevitably to automystification that provides spiritual comfort and a conviction of the strength of one's own position. Automys-tification deadens the sense of not only the truth, but also good, it sup-presses remorse, seeking adequate justification for the committed deeds. "Self-deception can satisfy: a. a desire to treat one's own actions as actions based on reasonable assumptions; b. a desire of moral justification of per-sonal choices; c. a need to maintain the coherence of views in all those cases where the truth threatens the attitudes" (ibid.), says Hirszowicz.

Janina and Zygmunt share the illusions of the epoch, which on the one hand formulates their choices, and on the other hand limits them. They, in different stages of life, fall for various 'promises' submitted by commu-nism. Among the latter, one should mention first of all satisfying the desire of belonging, political identification with something that goes beyond the routine of the daily life, identification with a 'higher matter,' usually asso-ciated with being in the community and the expression of radical opposi-tion to social inequality, uncontrolled mechanisms of capitalism, poignant poverty and humiliation. For Janina, who tries at all costs to break free from the nightmare of *alienated* childhood, the desire to do something mo-mentous after years of war vegetation and struggle for survival meant that communism, originally announcing the abolition of class divisions and ethnic minorities, finally turned out to be a great as much as an unfulfilled hope. "The Party never asks to be judged by the morality of its action. Its kingdom was no of this world. It beckoned to the future. And the fu-ture was all bliss—without hatred or prejudice, rice or nation. Was this not a world I had dreamed of ever since my life behind the ghetto walls? Was it not the only world in which, once and for all, my dream of belonging would came true?" she writes (J. Bauman, 1988, pp. 88-89; form as in the original).[4] Zygmunt, in turn, as a man brought up in a very poor family, experiencing firsthand the 'charms' of the Polish variety of capitalism and anti-Semitism, would not or rather could not distance himself from communism. While in the early development of his intellectual sensitivity, he understood Marxism primarily in terms of ownership and economic planning, with time, this ideology has become for him the only acceptable form of activity of the masses.[5] Marxism was an unmasking ideology, mobilizing people to resist against injustice, it was a kind of emancipation, giving the chance of liberation.

Hanna Świda (1997, p. 120) calls the people born between 1920 and 1930, the time when Bauman was born, the "spotty generation"—immature, ideologically empty youths who were given a chance to influence the reality, to be "creators of good faith." Maria Hirszowicz (2001, p. 113), in turn, lists a number of historical factors which facilitate understanding the "spotty generation": "'They were people living at the crossroads of epochs—their leftism was shaped, at least to a certain extent, by pre-war and wartime experiences. The memories of unbelievable poverty which was commonplace enough to make some parts of Poland look like the Third World, the resistance against anti-Semitism blossoming in pre-war Poland, the disappointment with the policies of the London Government and shock caused by the defeat of the Warsaw Uprising, being almost completely shut off from any information about what was really happening in Soviet Russia—all that favored the acceptance of Stalinist slogans praising the brave new world." In this atmosphere, in 1943, Bauman joined the Polish First Army and took part in liberating Poland. Immediately after the liberation, the 4th Jan Kiliński Division was reorganized to become the basis for the newly founded Internal Security Corps (KBW[6]). The fact that it was this and not any other division to become KBW was "a coincidence from the point of view of my life story… But there was no coincidence in the fact that I accepted this coincidence without a murmur," admits Bauman (Bauman, Kubicki & Zeidler-Janiszewska, 2009, p. 151)).

---

[4]   In the same tone as Janina Bauman, Aleksander Wat (1998, pp. 71-72) captures the atmosphere of that era in the following words: "And the warmth, the mutual love of this small cell surrounded by a hostile world, a strange one, it was a powerful glue."

[5]   This is particularly evident when one remembers Bauman's military period, his commitment to the activity of PPR, where Marxism was conceived primarily as an economic formation, then his fascination with the views of Stanisław Ossowski and Julian Hochfeld, associated with humanistic socialism (a fraction of PPS); these ideas are indeed still close to Bauman.

[6]   Korpus Bezpieczeństwa Wewnętrznego.

In the 'Trap of Assimilation'

Already as a soldier, political officer and a member of the Workers' Party, Zygmunt Bauman fell into the pitfall of assimilation, similar to the one into which there fell the assimilated Jews of the interwar period. The example of German Jews, cited in *Modernity and Ambivalence* shows that assimilation was—in some ways—the most serious threat that lurked also in the post-war Poland. The previously stigmatized Jews were deliberately allowed to anchor in the strong structure of the socialist state, with the hope that this will effectively weaken the collective Jewish identity. Bauman (1991, p. 107) writes, *inter alia*, "tolerant treatment of *individuals* was inextricably linked to intolerance aimed at collectivities, their ways of life, their values and, above all, their value-legitimating powers" (italics in the original). The autonomy of a beyond-individual identity of any non-socialist provenance was not acceptable for the authorities preparing to exert the totalitarian rule of power. However, the same authorities were willingly accepting to their ranks individual 'aliens' whose identity was measured in terms of their loyalty to the state. Since then, the world was not divided into Jews and non-Jews, but people ideologically mature and non-mature, or in other words: to those who understand the 'historical necessity' or are 'resistant' to it. The trap of assimilation was based on offering the assimilated persons equal access to power, which, through a somewhat circuitous route, generated informed and committed members of the strong structures of the system. On the other hand, those who belonged 'nowhere' were left with a strong sense of rejection, sometimes even taking the form of a kind of crippling loneliness.[7] The assimilated Jews always acted as if under the pressure to prove the ideological usefulness, paradoxically, however, the evidence of political involvement—just as in the world of pre-war Germany—could turn against them at any moment. "Without any conscious design on their part and without any noticeable pressures from outside, the Jewish activists of the socialist movement found themselves heavily concentrated in selected areas of party activity. They constituted a majority among party journalists, theorists and teachers of party schools. Those roles assured them of a central and highly prestigious role in party life, and through it in German politics as a whole. The same roles, however, made their position inside the party increasingly awkward and widely resented—the moment when the radical political movement of the early years ossified into a highly bureaucratized establishment [...]" (ibid., p. 147).

---

[7]   It is manifested, *inter alia*, in the following confession of Janina Bauman (1988, p. 15): "During the first year after the war I wanted to leave Poland—to get away from the place where I was seen as an unwanted stranger. I felt lonely at school, lonely among my neighbours, singled out and set apart in the very place where I felt I belonged."

Trying to make sense of the existence in the trap of assimilation, Bauman does it naturally in the scientific discourse, but in the depths of the soul, he is experiencing a personal historical paradox. *Modernity and Ambivalence* is a clear proof of the discovery in his own life story of an area fully taken over by strangeness. That is probably why in the Foreword for Polish readers we come across the following words: "This is a book about strangers. About how they became strangers. And how they tried to stop being them. And how they failed" (Bauman, 1995, p. 7), and further, in relation to the events of 1968, it is written there: "I could no longer come into public buildings—I would probably infect the walls, and certainly the people who stayed within them" (ibid., p. 9). In the same Foreword, Bauman emphasizes, which is highly meaningful, that his translator into Polish is his wife Janina, because "it rarely happens that the fate of the translator and the author resonate so closely. And it seldom happens that the translator and the author understand each other so well" (ibid.).[8]

Anti-Semitism and repressions related to it that Zygmunt Bauman experienced while living in Poland, after all, turned out to be less painful for him than the general disappointment with socialism. The system in which he saw himself as an engaged citizen made him an outcast, a stranger, an enemy of the people. Hence, after *Modernity and the Holocaust* there was a need for a book about how modernity (not only Nazism and capitalism, but also socialism) dealt with the problem of ambivalence.

Stefan Morawski would probably not agree with the thesis that Bauman himself fell into the trap of assimilation. He was not to be absorbed by the system. Morawski (1998, p. 30) points out that the author of *Postmodern Ethics* never changed his Jewish-sounding surname, because "he wanted to be himself, to challenge and test the new system which promised brotherhood rather than parochial xenophobia."

However, if we assume that Bauman stuck in the double trap (assimilation and commitment), this was not due only to anti-Semitism, poverty, and forms of humiliation provoked by it. The trap of commitment (and the consequent danger of automystification) was the greater, the stronger was the force of theoretical arguments, which (especially after the military period) he might have referred to more and more frequently. In 1948 he explained to his wife to be that "[…] there would be no room for anti-Semitism, or any other racial hatred, under Communism—this fairest of social system, which would guarantee full equality between human beings regardless of language, race, and creed. We were particularly lucky,

---

8   It is worth noting that already in the Polish period of Bauman's writing he referred to the experience of his wife. His work, *Wizje ludzkiego świata* (*Visions of the Human World*) is dedicated to "Janina, the companion of thoughts and actions."

he stressed, to have been born at the right time and in the right place to become active fighters for this noblest cause. The greatest of historical changes was happening before our eyes, here and now. To stand by as idle witnesses would be to miss a unique opportunity. Running way would be betrayal" (J. Bauman, 1988, p. 49; form as in the original).

Realizing that racial divisions did not disappear in the new system, and demobilized in 1953 due to his father's contacts with the Embassy of Israel, Bauman was suddenly left with no institutional support. Then again, submitting to the rules of inclusion and engagement he began to specify his belonging (if not loyalty) to the state. He found the socialistic grounding at the University of Warsaw. Only two years elapsed from his demobilization to his taking on a job as an assistant. On the one hand defending his life strategy and willingness to enter a strong structure, on the other the extraordinary power of theoretical grounds, pushed him into the abyss of self-deception: "In 1953, I still believed that communism could be led back to the right path... Whatever I found annoying and repulsive in the practices of 'the ruling powers,' I attributed to 'errors and distortions.' I saw more and more human misfortune, unjust judgments and disgraceful actions, but I failed to generalize them into a big picture. I did not think (or was afraid to believe?) that they were deliberate and fitting the newly introduced 'Polish system.' [...] My doubts did not concern the idea, but rather the way it was implemented," he explains (Bauman, Kubicki & Zeidler-Janiszewska, 2009, p. 163). Janina stresses that "since he had left the army he seemed to see things more clearly. Thought still a sincere socialist—which deep in his heart he has remained to this day—he now began to see that all was no right in this world of his. He became more and more aware of contradictions between word and deeds. Perhaps Marxist theory itself, conceived a century ago in different historic and social conditions, needed a new interpretation in our modern time and change society? Perhaps the duty of a communist who happened to be a scholar was to point his finger at what was wrong, to raise doubts, to rethink ideas which were clearly unworkable? Daily arguments which his university friends and colleagues helped to clarify his mind. His first critical essays followed. He was no longer a blind worshipper of the Party line" (J. Bauman, 1988, p. 115).[9] Thus, it is clear that Zygmunt Bauman has never ceased to be

---

[9] Jerzy Wiatr, a friend and a scientist involved in the activity of the Party, remembers that when in February 1956 a paper presented by Khrushchev revealed that Stalin had committed crimes, for many people it was a real shock. "Smart and honest people, one of my closest friends Zygmunt Bauman among them, were devastated. I was then struck by how our responses differed. For me, the disclosure (though only partial) of Stalin's crimes was something optimistic. Since I had known before that he had been guilty of crimes, I did not despair, but rather felt psychological liberation," writes Wiatr (2008, p. 40).

a Marxist, just as he has never entirely doubted in socialism, whose human face was to be an alternative for the individuals stuck in consumer lethargy. This belief, however, shaded him the truth about the everyday life in a socialist system.

The treachery of commitment is based on the conflict between individual aspirations (participation in the community, desire to belong somewhere, implementation of the principles of equality and freedom, fight against exploitation, poverty and humiliation) and the objectives that the Party actually implemented. Maria Hirszowicz (2001, p. 10) writes, "They do not remember how often the truth appeared within reach, but they looked away, instinctively guarding their faith. That faith was for many the defense of the chosen way of life; it was also a form of psychological adaptation to the new order, ensuring good mood and allowing to operate in the belief that there are higher goals that guide our actions." The belief that personal aspirations are reasonable and consistent with the ideological line of the Party results in a kind of moral torpor, and the very action in the Party gradually becomes an internalized need. At the same time, all the 'dangerous' practices of the Party are deemed to be mistakes that are sure to be noticed and timely corrected by the Party leadership. Communism, like any well-functioning, rational system, cleverly shifts the burden of moral responsibility from the consequences to the intentions of the perpetrators. And if the intentions were right, i.e. if they served the system, all the distortions, forgeries, and even crimes were passed and treated as a temporary inconvenience. It can be well seen in the confession of Janina who, describing the Party's imponderabilia in 1950, puts into Zygmunt's mouth an eloquent statement: "unfortunately [...] the Party ranks were still full of untrustworthy individuals, ruthlessly ambitious climbers and ideologically immature members. Yet, despite this transitory weakness, despite the grave mistake often committed in its name, the Party was the most powerful agent of social justice and had to be implicitly trusted. You cannot make an omelette, he said, without breaking eggs. You cannot make a revolution without accidentally hurting same of the innocent. The Soviet Revolution had created many victims and there had been many mistakes" (J. Bauman, 1988, p. 77).

How quickly the originally noble intentions can get warped knew no one who had not burned his hands. Again there worked the mechanism of automystification damping the moral instinct, which this time disappeared into the collective responsibility, in the common action: "Beware of those who promise you to bear responsibility of your deeds. What they want is your compliance. The responsibility will remain your own. Ignorance is not excuse for complicity. Do not let the powerful catch hold of your finger: they take your whole arm. And you will not even notice when they engulf the rest of you" (ibid., p. 89).

The ideological veil finally stopped masking the distortion of the system. Although Bauman, as I suggested above, remained a Marxist, step by step, he began to realize the gap between the ideal and the real world. "The hope that the 'party' will understand and admit their 'mistakes,' will turn back from the false path and [...] restore a human face to socialism, stubbornly remained then and even some time after the wiser and more keen than me, for example, Leszek Kołakowski, had reached the conclusion that it was not about the mistakes, but the system assumptions," he wrote elsewhere (Bauman, Kubicki & Zeidler-Janiszewska, 2009, p. 164). Revisionist views of Zygmunt Bauman emerged only in the early 1960s.[10] But, before that happened, he gave the political elites, like all engaged intellectuals at that time, the legitimacy of a policy of terror. He strengthened the regime by skillfully operating words. Therefore, if today something touches Bauman in a special way and if we demand accountability from him, it primarily refers to the responsibility for his words.

The key to examine the submission of intellectuals to ideology should not be sought in the label of an intellectual and his moral superiority over the common thinking, not in the validity of his choices, but in what Jean-François Revel (1989, p. 331) described as "an abundance of conceptual, logical, and verbal resources, which he uses to justify his choice." Although scientists do not have the superhuman ability to predict future events, nevertheless, they are characterized by skepticism and critical thinking skills that are often amplified by believing that the world could be better than it is at the moment. 'Innate' distrust towards what only appears to be clear promotes seeing the reality through the prism of human actions, intellectuals' actions that is. This effect was captured by Leszek Kołakowski (1978, p. 978) as "replacement of thought by engagement." The scientific view of the world provides the tools to enchant the world, which was brought by Bauman to "finally mighty words—which, as never before, were to become flesh..." (Bauman, Kubicki & Zeidler-Janiszewska, 2009, p. 147).

Bauman at some point implicitly trusted the Marxist theory, perfect in its totality and seduction, highly contributing to the conceptual resources, ensuring many 'mighty words.' The charm of the internal coherence,

---

[10] Jerzy Wiatr (2008, pp. 56-58) writes that "Bauman and Hirszowicz gradually shifted to a more radical position, which was probably influenced by the climate of the philosophy and sociology department of the Warsaw University, where Leszek Kołakowski played first fiddle and he was becoming more and more radical in his criticism of the post-October political situation."

a huge dose of rationality and logic of Marx's argumentation effectively choked the moral instinct, which is deeply and fully explained in *Postmodern Ethics*. The author made a long-awaited settlement with the past, giving valuable tips to the ascending generations.[11] Since that moment, he has continually referred to Stanisław Ossowski and Julian Hochfeld. In the first of them Bauman appreciated the effort of taking the responsibility for words. Towards the other he showed gratitude for gaining the awareness of the consequences of making any social action (including thinking). They made Zygmunt Bauman learn that sociology is a critical science, that it is not only *about* people, but also *for* people; thanks to them he—long before Mills—experienced a 'sociological enlightenment,' recognizing the urgent need to bind the fate of individuals with history in which they are embedded (cf. Bauman, Kubicki & Zeidler-Janiszewska, 2009; Bauman & Tester, 2003). The thought of Hochfeld and Ossowski will finally become a leaven of the inflected vision of sociology as a "companion in the diffi cult art of freedom" (Bauman, Kubicki & Zeidler-Janiszewska, 2009, p. 26). Remembering both mentors during his inaugural lecture on the occasion of taking over the Chair of Sociology at the University of Leeds, Bauman said: "More than ever we must beware of falling into the traps of fashion which may well prove much more detrimental than the malaise they claim to cure. Who knows, perhaps our vocation, after all these unromantic years, may become again a testfield of courage, consistency and loyalty to human values" (Bauman, 1972, p. 203). Maintaining the same conclusive tone, Maria Hirszowicz (2001, p. 11) says, however, that "the only effective way to safeguard oneself against ideological automystification is not so much distancing oneself from the current social problems, but keeping the constant awareness of the pitfalls of engagement and axiology, which determines the political choices—the rejection of the principle of making others happy against their will, sensitivity to everyday human affairs, respect for the individual, checking up on the representatives of power and opposition to all lies, regardless of the purpose for which they are used."

---

[11] Glimpses of a kind of 'reverse' thinking can be seen much earlier. In 1967, *Twórczość*, a journal of more literary than scientific character, published Bauman's article entitled *Notatki poza czasem* (*Notes Beyond Time*). Despite the fact that it is an essay in nature, its interpretation poses many difficulties. The author was highly critical in it of the scientistic tendency in the contemporary sociology and he was also severe in the assessment of modernity: "Cruel is the age of the worship of science and the death of god" (Bauman, 1967, p. 86). He also considers there the situation of individuals, pointing to the inability to solve the conflict between what is subjective in an individual and what makes this individual an object. Moreover, in the paragraphs devoted to love he seems to have already read Lévinas, which is not chronologically possible. This philosopher was discovered by Bauman only during his work on *Modernity and the Holocaust*.

The involvement in the life of the Party was for the Polish sociologists a proof of their willingness to participate in the creation of reality, involving the construction of a new social order. It was also an attempt to play a significant role in the process of social change (is it not what Charles Wright Mills called for, when he wrote about the promise made by sociology?).[12]

Love to Sociology

Since the 1960s, Zygmunt Bauman has sought to answer the question concerning the functions that sociology should perform. Publishing in 1964 a popular science book, *Socjologia na co dzień* (*Sociology for Everyday Life*) he offered the readers participation in the peculiar course of thinking about their own lives, because he wanted ordinary people to know what shapes their plans. Sociology in such form is not about the great problems of science, it does not explain where regimes come from, it rather refers to "the most personal matters—the matters of everyday life" (Bauman, 1964b, p. 8). Focusing the attention on things close to us all is a clear sign of cracks in the previously uniform structure of thought. Bauman begins to fear that the results of his earlier work are double-edged, useful not only for the society, but also and foremost used by the authorities. The book, *Wizje ludzkiego świata* (*Visions of the Human World*), published in the same year as *Socjologia na co dzień*, enhances the doubts as to how knowledge should be managed. Sociology is presented as a tool of social engineering, resorting to manipulation (personified by Parsons and Lundberg) and rationalization (in engaged sociology outlined by Mills and sociology in action presented by Gramsci). They are obviously still legislative visions of science (let us remember that *Legislators and Interpreters* will come out in print twenty years later),[13] but looking for inspiration in Mills, Gramsci,[14] Ossowski and Hochfeld, they care more for the experience of individuals than the mechanisms of the system practices.

---

[12] Sociologists in Poland (Bauman, Hirszowicz, Ossowski and Hochfeld) deeply believed that sociological reflection may change the world. And the degree to which this belief proved to be justified is shown by a certain anecdote told by Bauman: "During Mills's stay in Warsaw, Gomułka went on the radio to criticize an essay by my friend Leszek Kołakowski. We all trembled; having our fingers singed so many times before, we expected the worst. But Mills was elated: 'How lucky you are and happy you must be—the leader of the country responding to philosophical tracts! No one at the top pays any attention to what I am doing'" (Bauman & Tester, 2001, p. 28).

[13] Peter Beilharz (2000, p. 87) proposes to read *Legislators and Interpreters* as Bauman's confession of the former Marxist who still feels responsible for the world that as a legislator he co-created.

[14] It is worth noting that Bauman discovers and uses Gramsci to deconstruct Marxism long before his works are read by the English leftists.

Dennis Smith, wondering *Who is Zygmunt Bauman?*[15], makes a reference to the *First Letter to the Corinthians* by Paul the Apostle, suggesting that Bauman's intellectual biography reflects the three values, namely: faith, hope and love (cf. Smith, 1999, p. 33). Almost blind faith in socialism, which was to cure us of humiliation promoting universal equality and justice declined when—first in 1953 and then in 1968—the same socialism rejected Bauman due to his 'Jewishness.' The direct effects of anti-Semitic political action were manifested, among other things, in making him resign from the position of the head of the General Sociology Department, and then in removing him from the University of Warsaw. At the same time, the whole edition of the recently prepared book *Szkice z semiotycznej teorii kultury (Sketches of Semiotic Theory of Culture)*[16] was destroyed, and its author was forced to emigrate abroad. "Because of his origin, position in the scientific life in Poland, activity in the Polish United Workers' Party, identification with Marxism, Bauman became a particular object of offensive campaign organized by the media and politicians. His name was found in press releases used as the generic name, it was written with lowercase letters and in plural," reminds us Nina Kraśko (1995, p. 33). In spite of this, the author of *Postmodern Ethics* could still see the sense in the pursuit of a socialist utopia, perceiving in it the axiotic ground that does not permit forgetting about an individual. Marxism was/is a form of a struggle for culture, understood as an action exceeding human limitations.

Bauman watched closely and with great enthusiasm the labor movement in Poland, pursuing the ideas of Gramsci of working people aware of their situation, who, under an active civic attitude worth following effectively stand up for the right to the freedom of choice. The hopes placed in Marxism were therefore associated with the option of the birth of a civil society, rather than with the state apparatus. Bauman was convinced that 'Solidarity' being born in Poland in the late 1970s and early 1980s is a significant sign of maturation of socialism. At the same time, workers were protesting in the UK, there was a crisis of the system of production, and alienated individuals began to voice their opinions. Perhaps it is no coincidence that in such general mood a book *Memories of Class* (Bauman, 1982) was created—an expression of great hope for the final chance of socialism.

---

[15] This is the title of a chapter in Smith's *Zygmunt Bauman: Prophet of Postmodernity.*

[16] This work was a summary of Bauman's interest in the semiotic theory of culture, which he developed in the years 1966-1968. Within the institutional dimension, it led to the idea to bring into being, in 1967, the Department of Social Anthropology at the Faculty of Philosophy of the Warsaw University, and, at the turn of 1967/1968, also the Theoretical and Empirical Laboratory (see Tarkowska, 1995, p. 9-21).

When in the late 1980s *Freeedom* (Bauman, 1988) was published, the UK government of Margeret Thatcher had long been incorporating the privatization reforms, at the same time extremely efficiently pacifying the resistance of the working class. Bauman who was observing that situation got rid of the last illusions and gained the confidence that socialism ceased to be the counter-culture of modernity. Its place was taken by capitalism, greedily devouring its opposition. Thus, the old love for socialism and socialist-oriented sociology gradually took on the character of sentimental memories. Bauman's biographers noted in his room at the Leeds University two pictures; Peter Beilharz (2000, p. 79) noticed a picture of Marx standing on the bookshelf, but Richard Kilminster and Ian Varcoe (1996, p. 7) spotted a reproduction of Picasso's work hanging on the wall of the office and depicting Don Quixote. Both these figures seem to be helpful in the interpretation of the vision of Bauman's sociology, as well as the reconstruction of the life strategy adopted by him. From Marx, Zygmunt Bauman draws the belief that the world may be different—better than it is, and that we need a utopia to know which direction to follow. Utopia is a valuable sign-post, although it is known that it will never be fulfilled. Probably that is the reason why Don Quixote is also close to Bauman, with his irony, stubborn persistence of ambiguities, consistent eluding any routine solutions and negation of common sense. Marx is dedicated to the love of socialism, and the knight-errant disturbs the order of everyday life, thus making it more bearable.

Not taking any dictionary for the final one, Bauman, like the hero of the Cervantes' novel, is at times reminiscent of Richard Rorty—the liberal ironist. He has read many a novel and participated in more systems than one, he has reveled in the world in various forms, so now he can raise uncertainty, undermining the self-evident truths; he constantly makes others wonder, like Don Quixote who tells Sancho Panza to undermine the folk wisdom and common knowledge contained in the proverbs, which he continuously used. Where once he was convinced that he was in possession of all vital answers, now he is not even certain whether he asks the right questions. It is, therefore, human being in the world, so full of irony, that leads to a situation in which there will always be something else to be done. In addition to Marx and Don Quixote, I would place in Bauman's office also a picture of Alfred Schütz, both because of the immigrant life and thinking about sociology as an ironic strategy.

References

Bauman, J. (2011). *Nigdzie na ziemi: Powroty—Opowiadania.* Warszawa: Wydawnictwo Oficyna.

Bauman, J. (1988). *A Dream of Belonging: My Years in Postwar Poland.* London: Trafalgar Square Publishing.

Bauman, J. (1986). *Winter in the Morning: A Young Girl's Life in the Warsaw Ghetto and Beyond.* London: Free Press.

Bauman, Z. (1995). Do czytelnika polskiego. In Z. Bauman, *Wieloznaczność nowoczesna: Nowoczesność wieloznaczna (Modernity and Ambivalence*—Polish edition) (pp. 7-9). (J. Bauman, Trans.). Warszawa: Wydawnictwo Naukowe PWN.

Bauman, Z. (1993). *Postmodern Ethics.* Cambridge, MA: Basil Blackwell.

Bauman, Z. (1991). *Modernity and Ambivalence.* Cambridge: Polity Press.

Bauman, Z. (1990). *Thinking Sociologically.* Oxford: Basil Blackwell.

Bauman, Z. (1989). *Modernity and the Holocaust.* Cambridge: Polity Press.

Bauman, Z. (1988). *Freedom.* Milton Keynes: Open University Press.

Bauman, Z. (1987). *Legislators and Interpreters.* Cambridge: Polity Press.

Bauman, Z. (1982). *Memories of Class: The Pre-history and After-life of Class.* London: Taylor & Francis.

Bauman, Z. (1972). Culture, Values and Science of Society. *University of Leeds Review, 15,* 185-203.

Bauman, Z. (1967). Notatki poza czasem. *Twórczość, 10,* 77-89.

Bauman, Z. (1964a). *Wizje ludzkiego świata: Studia nad społeczną genezą i funkcją socjologii.* Warszawa: Książka i Wiedza.

Bauman, Z. (1964b). *Socjologia na co dzień.* Warszawa: Iskry.

Bauman, Z., & Tester, K. (2001). *Conversations with Zygmunt Bauman.* Cambridge: Polity Press.

Bauman, Z., Kubicki, R., & Zeidler-Janiszewska, A. (2009). *Życie w kontekstach: Rozmowy o tym, co za nami i o tym, co przed nami.* Warszawa: Wydawnictwa Akademickie i Profesjonalne.

Beilharz, P. (2000). *Zygmunt Bauman: Dialectic of Modernity.* London – Thousand Oaks – New Delhi: Sage.

Hirszowicz, M. (2001). *Pułapki zaangażowania: Intelektualiści w służbie komunizmu*. Warszawa: Wydawnictwo Naukowe Scholar.

Kilminster, R., & Varcoe, I. (1996). *Culture, Modernity and Revolution: Essays in Honour of Zygmunt Bauman*. London: Routledge.

Kołakowski, L. (1978). *Les intellectuels contr l'intellect: L'esprit réviolutionaire*. Bruxelles: Edition Complexe.

Kraśko, N. (1995). O socjologii zaangażowanej Zygmunta Baumana. In E. Tarkowska (Ed.), *Powroty i kontynuacje—Zygmuntowi Baumanowi w darze* (pp. 22-35). Warszawa: Wydawnictwo Instytutu Filozofii i Socjologii Polskiej Akademii Nauk.

Mills, C.W. (1959). *The Sociological Imagination*. Oxford: Oxford University Press.

Morawski, S. (1998). Bauman's Way of Seeing the World. *Theory, Culture and Society, 1* (15), 29-38.

Revel, J.-F. (1989). *La connaissance inutile*. Paris: Pluriel.

Sennett, R. (1977). *The Fall of Public Man*. New York: Knopf.

Smith, D. (1999). *Zygmunt Bauman: Prophet of Postmodernity*. Maiden, MA: Polity Press.

Świda, H. (1997). *Człowiek wewnętrznie zniewolony: Mechanizmy i konsekwencje minionej formacji—analiza psychosocjologiczna*. Warszawa: Zakład Moralności Aksjologii Ogólnej, Instytut Stosowanych Nauk Społecznych, Uniwersytet Warszawski.

Tarkowska, E. (1995). Koniec i początek, czyli próba antropologii społeczeństwa polskiego. In E. Tarkowska (Ed.), *Powroty i kontynuacje—Zygmuntowi Baumanowi w darze* (pp. 9-21). Warszawa: Wydawnictwo Instytutu Filozofii i Socjologii Polskiej Akademii Nauk.

Wat, A. (1998). *Mój wiek: Pamiętnik mówiony*, vol. 1-2. Warszawa: Czytelnik.

Wiatr, J. (2008). *Życie w ciekawych czasach*. Warszawa: Europejska Wyższa Szkoła Prawa i Administracji.

Chapter Five

MICHEL FOUCAULT AS A HETEROTOPIA[1]

by Marcin Kafar

Not all weighty humanistic thoughts are worth locating them in the bio-graphical subsoil, on which they grew; and in the case of others, the weave of their intertwinement with the author's life (and the resulting conse-quences) seems in turn so clear and significant at the same time, that it does not leave anyone indifferent, provoking to deeper reflection upon it. This latter category undoubtedly includes the achievements of one of the most intriguing (due to many co-occurring circumstances) 20th century thinkers—Michel Foucault. He himself, referring to his own work, con-fesses, "Whenever I tried to do some theoretical work, it was born of the elements of my existence, it always referred to the processes I've seen in my environment. Because it seemed to me that in the things that I see, in the institutions I am dealing with, in my relations to others I notice deep crevices, disruptions and dysfunctions—that's why I decided to take up such work, a kind of autobiographical fragment" (Foucault, 1981, as cited in Eribon, 2005, pp. 50-51). The fact that what the author of *Madness and Civilization* modestly refers to as the autobiographical fragment under-went multiplication over time, absorbing more and more areas of the un-usual life, is also confirmed by Didier Eribon skillfully moving around the

---

[1] This title, together with the idea it represents of looking at Michel Foucault from the heterotopic perspective was taken from the paper presented by Andrzej Paweł Wej-land, whom I would like to thank for sharing the paper and our inspiring conversations about heterotopicity in Foucault's works. The above-mentioned lecture (titled *Michel Foucault: Utopia and Heterotopia*), was given at the Department of Ethnology and Cul-tural Anthropology, University of Łódź in November 2004.

intricacies of Foucault's biography. Commenting *inter alia* on the sources of research on madness presented in *Histoire de la folie*, he says that "everyone, regardless of whether they knew the deeper causes of his disorders or not, remembered [Foucault] as a man who balanced on a rope with a sense of precarious balance that at any moment could slip towards madness. Also, in that fact, everyone sought the explanation of his obsession with psychology, psychoanalysis and psychiatry. 'He wanted to understand everything that was associated with privacy and exclusion,' says one of his colleagues. 'His very keen interest in psychology undoubtedly stemmed from his biography,' says another. We could also hear the confession: 'When *Histoire de la folie* was published, everyone who knew him clearly understood that this book is most closely connected with his personal story.' One of his then close friends said: 'I have always believed that one day he will write about sexuality. He had to give it the central place in his work, because it held the central place in his life,' and he continues: 'His most recent books present a sort of his personal ethics that he had fought for himself. Sartre never wrote ethics, Foucault did it,' or: 'Drawing on Ancient Greece, Foucault found in the *Histoire de la sexualité* archaeological uncovering of his own foundation…' In short, everyone agrees that the work of Foucault, and even his method, are rooted in a situation which he so dramatically experienced as a *normalien*. Of course, it's not mean that the whole work of Foucault should be interpreted from the perspective of his homosexuality […]. Obviously, with extreme simplification, one can observe how an intellectual project grew out of personal experience, which ought to be regarded as original, how a thinker prone to intellectual adventures 'invented' his idea throughout the confrontations of the personal and social life and did not stop at it, but began to think about it, to cross it, in order to—in the form of ironic rejection—ask it as a question to those who asked him before: 'Do you actually know who you are? Are you convinced about your common sense? Your scientific concepts? Your categories of perception?" (Eribon, 2005, pp. 49-50; italics in the original).

Since Foucault practiced—to follow his nomenclature—the endless *sobąpisanie (self-writing)*,[2] it seems appropriate to suppose that at his level of self-awareness (namely the methodological one), he did not do this blindfolded. Where, then, is the key to understanding the phenomenon

---

2   The phrase 'sobąpisanie' which is best reflected by the English 'self-writing,' I repeat after Michał Paweł Markowski, who translated in such a way the French phrase 'l'écriture de soi.' Markowski, aware of the importance of language nuances, in turn reveals that he followed this path, unable to find a more appropriate formulation for Foucauldian intuitions than the one based on the analogy of 'życiopisanie' ('life-writing') by a Polish poet Edward Stachura (cf. footnote 1 in Foucault, 1999, p. 304).

known as 'Michel Foucault'? What makes it? This matter, in accordance with the principle of narrative retardation, I will elucidate step-by-step, trusting in the Reader's patience. Let me start by giving the floor once again to Didier Eribon: "In connection with Foucault, there appeared—says his biographer—specific difficulties. Foucault was a complex and colorful personality. 'He wore masks and constantly changed them,' said Dumézil and he knew him better than anyone else. I did not try to uncover 'the truth' about Foucault: behind each mask another mask appeared, and I do not believe in the possibility of extracting the truth about the personality emerging from one shell after another. Could there be a number of Foucaults? Thousands of Foucaults, as Dumézil stated? Undoubtedly" (ibid., p. 10). It seems that despite the pessimism, which is concealed in these words, at least at the first glance, the author, following Georges Dumézil, reveals a 'truth,' and indeed an important truth nailing, in my opinion, the personality-identity complexities of Foucault and multiple effects resulting from them. Who was Foucault? Did anyone, including himself, know the answer to it? Finally, to what extent, and for who, did it matter? Quite meaningful—perhaps not only in a symbolic sense—are in this regard the trials and tribulations of naming the newborn boy: "[...] the family gave the son the name of his grandfather and his father; Paul—grandfather Paul Foucault, Paul—father Paul Foucault, Paul—the son of Paul Foucault... But Mrs. Foucault did not want quite to conform to the tradition of her husband's family. Yes, her son was to be named 'Paul.' So be it. But she added the middle name, 'Michel,' and joined both of them with a hyphen. The official documents and school reports contained the name 'Paul.' End of story. However, the family soon started using the other name: just 'Michel.' For Mrs. Foucault he would forever remain Paul-Michael, she remembered him with both these names just before her death. The whole family to this day speaks of 'Paul-Michel.' Why did Foucault change his name? 'Because his initials—P.-M.F.,' said Mrs. Foucault, 'were the same as Pierre Mendes France's initials.' This explanation was provided to her by her son. However, he presented the matter quite differently to his friends: he did not want to have the name of his father, whom he hated in his youth" (ibid., p. 19).

Stubborn adherence to tradition, fighting an uneven battle with the juggling of appearances, skillful dodging, instead of facing the name-related norm imposing lasting social obligations—that is Foucault's fate, which, it is necessary to add, was taken by him with a slightly ironic smile, though not free of resignation, blooming on the face of the genius alienated from the world. "This is the city where I was born; decapitated saints with books in their hands ensure that justice is infallible, the castles are armed... this is the hereditary gift of my wisdom" (ibid., p. 18), Foucault says about Poitiers, a city where he spent his childhood and early adolescence; does he say it as 'Paul,' 'Michel,' or 'Paul-Michel'? Or maybe a little bit as each of them? An intellectual floating on high seas; a provincial returning after many years to his homeland; a sensitive man maintaining strong ties with the mother till the end of his days; a rebel bearing a deep grudge against the father... Undoubtedly, no matter what perspective it is set into, the (self-)portrait of Foucault is painted in different colors, none of which is strong enough to pierce through the rest of them...

This multicolored nature, deliberately, I suppose, reinforced by Paul-Michel himself, manifested outside in a wide repertoire of bizarre behaviors that, sooner or later, made Foucault, almost everywhere where he appeared, an unwanted person. "He was an incorrigible individualist, and his relationships with others were complicated, and sometimes swollen with conflicts. He did not feel good in the new skin—there was something morbid about it. [...] Foucault hid in his own solitude, he abandoned it only when he could mock others. He ridiculed his colleagues with such coldness that soon he became known for it. Mocking and jibing, he gave them blunt nicknames and used them in public [...]. He argued with everybody, he made enemies everywhere, showing the signs of odd aggressiveness, sometimes going hand in hand with truly outstanding megalomania. [...] As a result, he soon became universally hated" (ibid., pp. 46-47)—such were the words used by Eribon to describe the situation in which the future philosopher found himself right after being admitted to the elite of École Normale Supérieure in 1946. Over the time, the atmosphere around Foucault thickened until the moment when, for the sake of the good name of the noble institution and the mental health of the eccentric young man, it was decided to take radical steps towards him: "One day, a teacher found him lying on the floor in the classroom—he saw that the boy cut his skin on his chest with a razor; another time he was seen chasing a student with a dagger in his hand. And when in 1948 he attempted to commit suicide, most colleagues saw in this a confirmation of what had long been assumed: Michel Foucault's mental balance was in a condition worse than deplorable. [...] Two years after Foucault had been admitted to École he found himself in St Ann's Hospital—where he was entrusted to the care of Professor Delay, one of the leading French psychiatrists. [...] This was the first encounter of Paul-Michel with psychiatry as an institution. It was also the first close-up to that uncharted border which—perhaps not as dramatically as it is taken—separates a 'madman' from a 'sane man.' In any case, this painful episode gave Foucault a privilege, which was widely envied: he got a single room nearby the school clinic. In this way, he could isolate himself and gain peace of mind that he needed to work. He lived there again when in the years 1950-1951 he was preparing for the second time for aggregation; and he appeared there one more time when he was giving lectures—but this time because of mere the convenience. Meanwhile, he yet again attempted to commit suicide, or in any event he staged such trials. 'Foucault was quite obsessed with this idea,' says one of his friends. One day, when someone asked, 'Where are you going?' Foucault replied, 'To BHV, I want to buy rope, with which I will hang myself.' The school doctor, obliged with medical confidentiality, merely stated, 'These disorders resulted from the improper attitude toward his own homosexuality.' Indeed, whenever Foucault returned from his frequent nocturnal visits to gay bars and pickets, he was depressed for hours, sick, and sullen with shame [...]" (ibid., pp. 47-48).[3]

'Incompatibility' of Michel Foucault resulted in his falling into deep loneliness that could be relieved, at least in part, by the act of writing—as believed the author of *La Volonté de savoir*: "Do you know why a man writes?" he once asked his assistant, Francine Pariente, and he answered this question himself, "To feel loved"…

Paul-Michel desired dialogical engagement with the audience coming in great numbers to his lectures given at the Collège de France, almost as much as he desired genuine love (the taste of which he apparently experienced a number of times not only in the text dimension of life). Unfortunately, each time, instead of questions stimulating the discussion, there came big disappointment: "What I said here, needs to be discussed," he explained, adding, "Sometimes, when a lecture is not going so well, one question is enough to put it on the right track. But the question is never asked. In France, the group effect prevents any real discussion. And since the lecturer does not receive any feedback, the lecture becomes more and

---

[3] More than once his homosexuality brought Foucault to fight for a place (social one), just as he fought in the late 40s and early 50s, after he had been awarded a scholarship of the Thiers Foundation. Eribon says, "Foucault took benefit from the scholarship only for a year, instead of—as provided in the statute—for three years. He found life in the community difficult to endure [...]. Each scholarship holder had his own room, and therefore he could enjoy relative freedom, but in spite of it all, the scholarship holders lived in the guest house where they had to cope with life in a group of about twenty people, because—apart from the holders of 1951—there lived scholarship holders from the previous years. All meals were consumed in the same company. Foucault once again managed to alienate himself from everyone. He made attacks on each of them, pulling faces and inciting fights. His relations with his colleagues were marked by constant conflicts—the situation ended with a drama, because Foucault established an affair with one of his colleagues and the whole story ended grimly, because he was suspected of taking over letters… He did not want to go on with it all any longer, and his colleagues did not keep him either" (Eribon, 2005, p. 65).

more theatrical. Towards the people sitting in the audience I behave like an actor, or an acrobat. And when I come to the end of the lecture, I am struck by the feeling of complete loneliness" (ibid., p. 273)…

∞

Famous exegetes of Foucault's works, Charles C. Lemert and Garth Gillan, prompt—but they do it for the sake of other needs, relatively different from mine—that to understand Michel Foucault's thought it is useful to apply "his own terms" (Lemert & Gillan, 1982, p. XV). This unusual procedure, that is supposed to protect us from falling into the trap of misinterpretation of the complex work of the author of *Les mots et les choses*, seems to work out also in the area of exploring mutual interactions between the texts and life, *the* texts and *the* life. I have come to the similar conclusion dealing with the wealth of Foucault's metaphors, which include the one that strongly merges the 'professional' and 'non-professional' dimensions of philosopher's actions. I mean here the concept of **heteroclicity**, first used in the Preface to *The Order of Things*. From there, we learn that the book, which once became so stunningly popular "arouse out of a passage in Borges, out of the laugher that shattered, as I read the passage, all the familiar landmarks of my thought—*our* thought, the thought that bears the stamp of our age and our geography—breaking up all the ordered surfaces and all the planes with which we are accustomed to tame the wild profusion of existing things, and continuing long afterwards to disturb and threaten with collapse our age-old distinction between the Same and the Other. This passage quotes a 'certain Chinese encyclopedia' in which it is written that 'animals are divided into: (a) belonging to the Emperor, (b) embalmed, (c) tame, (d) sucking pigs, (e) sirens, (f) fabulous, (g) stray dogs, (h) included in the present classification, (i) frenzied, (j) innumerable, (k) drawn with a very fine camelhair brush, (l) *et cetera*, (m) having just broken the water pitcher, (n) that from a long way off look like flies.' In the wonderment of this taxonomy, the thing we apprehend in one great leap, the thing that, by means of the fable, is demonstrated as the exotic charm of another system of thought, is the limitation of our own, the stark impossibility of thinking *that*" (Foucault, 1994, p. XV; italics in the original). Foucault explains then, reaching for suitable analogies and comparisons clarifying his reasoning, what is included in that *impossibility of thinking* as he puts it. Its meaning exceeds the plain and simple existence of the fabulous animals and the animals belonging to the remaining 'sophisticated' categories; it's not about the astounding combination, either, but about **the place** bringing them to **joined existence**, about *tópos koinós—common place*, whose peculiarity is compared to the autopsy table, on which there suddenly land the sewing-machine together with the umbrella; about the place creating an opportunity for a localization, though an extremely unlikely one, of a sudden gathering of all "worms and snakes" in the mouth of Pantagruel's companion, Eusthenes. Only there, paradoxically (!)—exclusively there in "that welcoming and voracious mouth" they are provided "with a feasible lodging, a roof under which to coexist" (ibid., p. XVI).

Obviously, the example of the alleged Chinese encyclopedia should be perceived just as an excuse to discuss much more serious issues relating to the basics of certain cultural processes. Without going into too much detail of Foucault's arguments, it is worth mentioning that they deal with the contact point of spaces, words and things. Culturally situated things and the corresponding words are in advance assigned to certain spaces, operating according to the principle of the "mute ground" where "it is possible for entities to be juxtaposed" (ibid., p. XVII) (the original context locations of the umbrella and the sewing-machine are maintained, based on the universally adopted classification canon, in the extreme distance from each other; thus it is usually much easier for us to think of them as objects of 'practical' use than, for instance, objects stimulating imagination). In Foucauldian logic, the subject under discursive pressures, constantly revises its classification actions as for the—largely unrealized—"fundamental codes" imposing the "empirical orders" onto the world (ibid., p. XX). Yet, what the interpreter is guided at by the specifically understood Borges' list, is the fact that there are situations where the empirical orders are sometimes questioned, their rules are doubted, and the one who, zounds (!), enters the orbit of the impact of this 'aberration' will feel deep anxiety and will become at the same time exposed to the "disintegration of language"—and will be threatened by the "loss of what is 'common' to place and name" (ibid., p. XIX). "It appears that certain aphasiacs, when shown various differently coloured skeins of wool on a table top, are consistently unable to arrange them into any coherent pattern; as though that simple rectangle were unable to serve in their case as a homogeneous and neutral space in which things could be placed so as to display at the same time the continuous order of their identities or differences as well as the semantic field of their denomination. Within this simple space in which things are normally arranged and given names, the aphasiac will create a multiplicity of tiny, fragmented regions in which nameless resemblances agglutinate things into unconnected islets; in one corner, they will place the lightest-coloured skeins, in another the red ones, somewhere else those that are softest in texture, in yet another place the longest, or those that have a tinge of purple or those that have been wound up into a ball. But no sooner have they been adumbrated than all these groupings dissolve again, for the field of identity that sustains them, however limited it may be, is still too wide not to be unstable; and so the sick mind continues to infinity, creating groups then dispersing them again, heaping up diverse similarities, destroying those that seem clearest, splitting up things that are identical, superimposing different criteria, frenziedly beginning all over again, becoming more and more disturbed, and teetering finally on the brink of anxiety" (ibid., p. XVIII).

Foucault strongly stresses the importance of these, let's call them—'altered states,' seeing in them a thoroughly ambivalent force, that with the same easiness can *destroy* the original order and *initiate* the creative cultural movement; this force sprouts in the **'dimension of heteroclicity,'** where "a large number of possible orders" constantly flash and transform into two opposing forms: **utopias** and **heterotopias**. "*Utopias* afford consolation: although they have no real locality there is nevertheless a fantastic, untroubled region in which they are able to unfold; they open up cities with vast avenues, superbly planted gardens, countries where life is easy, even though the road to them is chimerical. *Heterotopias* are disturbing, probably because they secretly undermine language, because they make it impossible to name this *and* that, because they shatter or tangle common names, because they destroy 'syntax' in advance, and not only the syntax with which we construct sentences but also that less apparent syntax which causes words and things (next to and also opposite one another) to 'hold together.' This is why utopias permit fables and discourse: they run with the very grain of language and are part of the fundamental dimension of the *fabula*; heterotopias (such as those to be found so often in Borges) desiccate speech, stop words in their tracks, contest the very possibility of grammar at its source; they dissolve our myths and sterilize the lyricism of our sentences" (ibid., p. XVIII; italics in the original).

Heteroclicity is revealed through language (to that extent it is recognized in *The Order of Things*), but it also goes—as shown by the considerations contained in *Des espaces autres*[4]—far beyond it; utopias and heterotopias represent a kind of a reservoir of *spatial possibilities* gaining universal resonance, since they can be found in each type of culture. They make an integral part of social activities stretched between the desire to idealize the world (utopia), and doubting in its legitimacy (heterotopia). While heterotopias remain peculiar spaces, they also have the value of a kind of

---

[4] This is the original title of the lecture given by Michel Foucault, March 14, 1967. The opportunity to present the concept of heterotopia was provided by a conference of Cercle d'Etudes Architecturales. The text of the speech was published in 1984 in the French journal *Architecture, Mouvement, Continuité*.

"counter-sites," that is places where "**all the other real sites** that can be found within the culture, **are simultaneously represented, contested, and inverted**" (Foucault, 1984a; bold added by M.K.).

∽

What I have said so far is enough to return to the problem of Foucauldian 'I', this time merging it directly with the idea of heteroclicity. Michel Foucault, *as a person* and *as a thinker*, is in my opinion 'another space'; remaining someone extremely **heteroclitic**,[5] he is at the same time a **heterotopia** and a **heterotopologist**[6]—someone who has, to use a metaphor again, the attributes of Eusthenes and simultaneously is aware of his own internal contradiction, which he ultimately transforms into the intellectual and cognitive advantage.

After these explanations it is easy to see that the 'taxonomic' trouble so meticulously enumerated by Eribon and Dumézil ("He wore masks and constantly changed them," "I did not try to uncover 'the truth' about Foucault: behind each mask another mask appeared," "Could there be a number of Foucaults? Thousands of Foucaults?") is immediately solved by its location in the common place, at *this spot*, remaining beyond all *other real places*. Significantly, the tone of the voices cited above is repetitive and resonant; the same Dumézil states, for example, "Foucault's intelligence was literally boundless, it was even *sophisticated*. He extended the field of his observations to embrace all areas of the revived being, where the traditional divisions into body and spirit, drive and ideas seem absurd: madness, sexuality, crime. Hence, his eyes like a spotlight swept the past and the present, ready to make even the most disturbing discoveries, and ready to accept anything, except for the requirement of keeping orthodoxy. It was the intelligence of many focal lengths, using

---

5   *Webster's Dictionary* has two alternative terms corresponding to the Polish adjective 'heterokliczny.' The first includes 'heteroclitic' or 'heteroclitical,' while the second is 'heteroclite.' These words (appearing in the literature already in the first half of the eighteenth century) reflect well the specific personality of Foucault, meaning as much as "deviating from ordinary forms or rules; irregular; anomalous; abnormal." The noun 'heteroclite' can also be used to describe a person who has an eccentric and unconventional way of behavior (*Webster's...*, entries: heteroclitic; heteroclite).

6   Talking about a 'heterotopologist' I follow the suggestion of Foucault, who, wondering how to describe heterotopias, proposes to elaborate "a sort of systematic description [...] that would, in a given society, take as its object the study, analysis, description, and 'reading' [...] of these different spaces, of these other places. As a sort of simultaneously mythic and real contestation of the space in which we live, this description could be called **heterotopology**" (Foucault, 1984a; bold added by M.K.).

moving mirrors,[7] replicating the formed judgments to turn them at once into their opposites, without leading to self-destruction, and not giving up" (Eribon, 2005, p. 412; italics in the original). Clifford Geertz, on the other hand, while preparing in 1978 a review for the English edition of *Discipline and Punish*, already at the very beginning attempted to equal Foucault-the-thinker with the non-places by Maurits Cornelis Escher. Since *Folie et déraison* was printed, its author, as declared by the eminent anthropologist, "has become […] a kind of impossible object: a non-historical historian, an anti-humanistic human scientist, and a counter-structuralist structuralist. If we add to this his tense, impacted prose style, which manages to seem imperious and doubt-ridden at the same time, and a method which supports sweeping summary with eccentric detail, the resemblance of his work to an Escher drawing—stairs rising to platforms lower than themselves, doors leading outside that bring you back inside is complete" (Geertz, 1978). Stephen J. Ball (1990, p. 1) notes that "Michel Foucault is an enigma, a massively influential intellectual who steadfastly refused to align himself with any of the major traditions of western social thought. His primary concern with the history of scientific thought, the development of technologies of power and domination, and the arbitrariness of modern social institutions speak to but stand outside the main currents of Weberian and Marxist scholarship. In an interview in 1982, in response to a question about his intellectual identity, Foucault characteristically replied: 'I don't feel it is necessary to know exactly what I am. The main interest in life and work is to become someone else you were not in the beginning. If you knew when you began

---

7 The 'moving mirrors' metaphor should arouse double curiosity, since a mirror, as such, places itself exactly at the intersection of utopia and heterotopia. Foucault (1984a) writes, "I believe that between utopias and these quite other sites, these heterotopias, there might be a sort of mixed, joint experience, which would be the mirror. The mirror is, after all, a utopia, since it is a placeless place. In the mirror, I see myself there where I am not, in an unreal, virtual space that opens up behind the surface; I am over there, there where I am not, a sort of shadow that gives my own visibility to myself, that enables me to see myself there where I am absent: such is the utopia of the mirror. But it is also a heterotopia in so far as the mirror does exist in reality, where it exerts a sort of counteraction on the position that I occupy. From the standpoint of the mirror I discover my absence from the place where I am since I see myself over there. Starting from this gaze that is, as it were, directed toward me, from the ground of this virtual space that is on the other side of the glass, I come back toward myself; I begin again to direct my eyes toward myself and to reconstitute myself there where I am. The mirror functions as a heterotopia in this respect: it makes this place that I occupy at the moment when I look at myself in the glass at once absolutely real, connected with all the space that surrounds it, and absolutely unreal, since in order to be perceived it has to pass through this virtual point which is over there."

a book what you would say at the end, do you think that you would have the courage to write it?"[8] Interestingly, the similarly self-reflective conclusions are, as it turns out, just a repetition of those from the early 1960s, when Foucault (1972, p. 12) began to see, paraphrasing his own words, *otherness* within his *own* thoughts. This takes place in the work deeply questioning the methods, borders and guiding clues relevant to the history of ideas. At the end of the chapter introducing us into the subject of archeology of knowledge, we find an intriguing internal-external dialogue, which is marked in the climax with a metaphor of a (text) maze, where the author—as we are convinced by the rhetoric used there—carefully hid his face. While making a momentary "autobiographical pact"[9] with himself, Michel-trickster poses a question, this time acting from the position of his *alter ego*: "'Aren't you sure of what you're saying? Are you going to change yet again, shift your position according to the questions that are put to you, and say that the objections are not really directed at the place from which you are speaking? Are you going to declare yet again that you have never been what you have been reproached with being?" and then states in a fickle way: "Are you already preparing the way out that will enable you in your next book to spring up somewhere else and declare as you're now doing: no, no, I'm not where you are lying in wait for me, but over here, laughing at you?'

'What, do you imagine that I would take so much trouble and so much pleasure in writing, do you think that I would keep so persistently to my task, if I were not preparing—with a rather shaky hand—a labyrinth into which I can venture, in which I can move my discourse, opening up underground passages, forcing it to go far from itself, finding overhangs that reduce and deform its itinerary, in which I can lose myself and appear at last to eyes that I will never have to meet again. I am no doubt not the only one who writes in order to have no face. **Do not ask who I am and do not ask me to remain the same**: leave it to our bureaucrats and our police to see that our papers are in order. At least spare us their mortality when we write'" (ibid., p. 17; bold added by M.K.).

This passage is clearly separated in the original (with double spacing) from the rest of the argument; and thus the *content* is symbolically

---

[8]   The phrase cited by Ball is taken from an interview conducted by Rux Martin during a seminar at the University of Vermont visited by Foucault in 1982; see Martin (1988).

[9]   The term 'autobiographical pact' is used by me according to the intentions assigned to it by Philippe Lejeune, assuming that we can talk about an autobiography from the moment the condition is fulfilled concerning the coexistence of a trivalent person in the text: the author, who is also the main character in the story and an actually existing person (cf. Lejeune, 1989).

replaced with a sign of *broken continuity*, and it is an attempt—when we again adopt the Foucault's perspective—to designate "**this blank space** from which I speak, and which is slowly taking shape in a discourse that I still feel to be so precarious and so unsure" (ibid.; bold added by M.K.).

The examples of *personality* and *mental* 'flashes' of Foucault's heterotopias could be multiplied within the area of their multi-layer and multi-aspect character, nevertheless, a matter equally important to looking for the evidence of the **heteroclicity principle** is a reflection on the effects that it produces; and these are quite significant.

We already know that heterotopias raise concern, putting some of us on the edge of fear, but is it so that the only persons that are saturated with this primal fear are 'aphasiacs' embraced by the all-powerful impotence of speech, sensation, movement or memory? Indeed, they are bothered by the lack of the right word or sound, but also— which is equally important due to the stigmatizing special kind of 'loss,' they often become a big trouble for the 'non-aphasiacs,' because they are not sufficiently 'translatable.' Michel Foucault reaches the limits of 'translatability' from the other side, he is, so to speak, '**hyper-expressive**'—he gives at the same time the impression of being 'nowhere' (in the blank space) and 'everywhere' (in all spaces at once); "None the less, his work has been taken up or has impacted upon a wide range of disciplines—sociology, psychology, philosophy, politics, linguistics, cultural studies, literary theory," says Stephen J. Ball (1990, p. 1), eagerly supported by Leszek Koczanowicz (1999, p. 7), emphasizing that considering the "whole multiplicity of reception and interpretation of the thoughts coined by the author of *History of Sexuality*, there is a consensus about the fact that his work has revolutionized philosophy and social sciences. Even a cursory review of the bibliography published in 1983 that enumerated Foucault's works and papers dedicated to him shows the scale of the impact.[10] It reports 2,570 works in different languages and on almost all disciplines in the humanities and social sciences. Since that time, the number of manuscripts devoted to Foucault's views as well as knowledge of his work have grown quite rapidly." These comments should be supplemented by the opinion, which to my mind is not an isolated one, expressed by Charles C. Lemert and Garth Gillan (1982, p. XIII), that Foucault "Not only [...] does write across disciplinary

---

[10] There are many indications that Foucault was aware of the repercussions caused by his texts. To the criticism of *La Volonté de savoir*, presented by Jean Baudrillard, proclaiming that Foucauldian work is imminently bound to be forgotten, its author sarcastically retorted: "My problem would rather be to remember Baudrillard," "It's enough for any scribbler to add his name to mine and whoever he may be, he can be sure of commercial success" (Eribon, 2005, p. 340).

boundaries, he is also read in this way. In other words, Foucault's readers tend to be those who, to greater or lesser extents, acknowledge that no single intellectual specialization is sufficient to the task of explaining the social world. They are, it seems fair to assume, people who share Foucault's conviction that the disciplines are both insufficient and part of the problem of modern society."

Observations on the extraordinary popularity of Michel Foucault and the fact that his work appears in numerous theoretical, analytical and interpretation contexts, (including those that go beyond the purely academic area), are completed and also explained by the **dispersion** category proposed by Umberto Eco. Trying to figure out the causes of worship that selected works enjoy, the Italian researcher was tempted to hypothesize that at the root of this behavior—on the part of the recipients—there is the 'fragmented' nature of these works. The author of *Name of the Rose* is of the opinion that the model dispersion compositions include *Casablanca*, Shakespeare's *Hamlet* and the Bible. Dispersibility assumes the opportunity of reversing something, giving it a new form, and finally—playing a perverse game of meanings consciously or unconsciously provoked by the author and undertaken by the recipient. "It is widely known—writes Eco (2007, pp. 157-158)—that *Casablanca* was filmed day after day without knowing the end of the film" ["If you knew when you began a book what you would say at the end, do you think that you would have the courage to write it?" wonders Foucault; note added by M.K.]. [...] Ingrid Bergman looks so charmingly mysterious in the movie, since playing her role she did not know which of the men she would ultimately choose, both received her tender and ambiguous smiles. We also know that in order to induce the dynamics of the plot, the script writers placed all the hackneyed tricks taken from the history of film and novel, turning the film into a museum [...] for moviegoers. Therefore, this movie can be seen as a set of archetypes. To some extent, the same can be said about *The Rocky Horror Picture Show*, which is a *par excellence* cult film, due to the fact that it lacks a form, so it can be continuously deformed and turned inside-out. [...]

The Bible owes a lot of its huge and centuries-lasting popularity to its dispersive structure, which comes from the fact that the Bible was written by a number of different authors. *Divine Comedy* does not have a disperse nature, but because of its complexity, because of the large number of characters and events appearing in it (everything to do with Heaven and Earth, said Dante), each of its verse can be distorted, exploited as a magic spell or as a memory exercise. Some fanatics have gone so far that they have used the poem in social games. The work by Dante, similarly to Virgil's Aeneid, functioned in the Middle Ages as a manual of fortune-telling, just like the *Centuries* by Nostradamus (another example of the success achieved through radical, incurable dispersibility). But, although *Divine Comedy* can be turned inside-out, we cannot do this with *Decameron*, because every story should be treated as a whole. The extent to which a specific work of literature can be subjected to turning inside-out does not depend on its esthetic values. *Hamlet* is still a fascinating drama (and Eliot himself cannot convince us to like it any less), I do not believe, however, that even the greatest fans of *The Rocky Horror* would attribute to it Shakespearean greatness. On the other hand, both Hamlet and *The Rocky Horror* make an object of worship, because their form is 'susceptible to dispersion,' and the latter is so dispersive that it allows us a variety of interactive games. To become a holy Forest, the forest must be convoluted and complex as the forests of druids and not structured like French-style gardens."

The dispersibility of Foucault's texts somewhat overlays their heterotopicity. This can be observed in the scope of the undertaken topics,[11] and the way they are presented; the subtlety, or even the vertiginous refinement of the paths of thought followed by the author of *Madness and Civilization*, allowed him to see the passages (sometimes relatively safe, but more often breakneck and by all means treacherous), where others faced a solid wall or, even worse, a bottomless abyss. His innate curiosity and exploration courage, both supported by a large dose of creative troublemaking, constantly pushed him into areas where he could ruthlessly expose ideological and ethical weaknesses underlying certain *discursive systems*. Foucault undoubtedly developed *blank spaces* (let's remember that he himself was a peculiar essence of the latter), but he always proceeded to them from *the middle of the developed spaces*; beginning with a critique of what was ambient (literally and figuratively), to determine then the indirect pathway, usually provided with a number of side routes. In this way, he prepared the ground for other alternative solutions, though immediately putting an emphasis on their temporality and the positively indexed (!) vulnerability to future change inscribed in it: "I wouldn't want what I may have said or written to be seen as laying any claims to totality. I don't try to universalize what I say; conversely, what I don't say isn't meant to be thereby disqualified as being of no importance. **My work takes place between unfinished abutments and anticipatory strings of dots. I like**

---

[11] Jerzy Szacki (2002, p. 902), admitting that Foucault "undoubtedly belonged to the original and influential thinkers of the second half of the twentieth century," notes as well that the author of *The History of Sexuality*, was "also one of the thinkers of his time, whose work led to a wide public resonance. It happened so that the works of Foucault [...] **touched issues vital not only for professionals—such as hospital, mental illness, prison system, human sexuality or the ubiquity of power**" (bold added by M.K.).

to open up a space of research, try it out, and then if it doesn't work, try again somewhere else. On many points—I am thinking especially of the relations between dialectics, genealogy and strategy—I am still working and don't yet know whether I am going to get anywhere. What I say ought to be taken as 'propositions', 'game openings' where those who may be interested are invited to join in; they are not meant as dogmatic assertions that have to be taken or left en bloc. My books aren't treatises in philosophy or studies of history: at most, they are philosophical fragments put to work in a historical field of problems" (Foucault, 1987, p. 100; bold added by M.K.).

CR

The example of Michel Foucault—this *another space, which is home to all the other spaces*, distinctly reminds us of the **complexity of the mechanisms that lead to the emergence of new ideas (in science)**. Here again the question returns, to which giving the answer was cleverly evaded by Ludwik Fleck, namely, what is the place of individuals in the thought innovation process? Are they just a background for the complex group activities, where the relationships occurring between knowledge and power minimize the efficiency of the work of a conscious human subject, or does the latter, as indicated by the theme of biographical epiphanies entwined with intimate experiences (see Chapter Two in this book), play a more serious role in the cited context? Should, in turn, we accept the thesis of at least relative symmetry that exists between what is individual and what is collective in the thought innovation process, what specific elements determine the latter? Dilemmas close to mine probably bothered also a Polish philosopher Jan Dembowski, who wrote many years ago: "Just as the thinking of an individual is not independent, but dependent on the collective, the given collectives cannot be understood as independent units. It is not difficult to quote many examples of how the social style of thinking affects the work of a biologist, and how natural theories depend on the religious views of their authors or their mathematical education. Collectives intermingle, and although a man belongs to many collectives at a time, he always remains the same person, he is a uniform man. His behavior is consistent with the molded nature and cannot be split. In all collectives, the way of thinking of an individual remains the same, and only the thinking material is different, which leads to different effect. **It is not clear [...] what the origin is of new ideas in science. Are we entitled to attributing the entire content of the human psyche to outside influences, completely excluding the endogenous factors? It seems to me that we are not. It can be thought that there are 'mutations,' which in the mind** of an individual provide the background for completely new associations independent from the influence of the collective. In other words, in some cases, a thought collective can be replaced by an individual that becomes the source and center of the new collective. Therefore, in this case, **individual creativity is possible in science**" (Dembowski, 1939, p. 439; bold added by M.K.).

The case of Foucault, in my opinion, is one of those exceptional cases *through* which, and *thanks to* which, ideas gain great and unexpected momentum (becoming "transdiscursive" as Foucault (1984b) would say), the source, if not the essence of which is a particular individual (*persona*) thrown into the dramas of life, who is also someone constantly struggling with internal constraints, drawing from them the power to incessant *becoming oneself as The Other*.

References

Ball, S.J. (1990). Introducing Monsieur Foucault. In S.J. Ball (Ed.), *Foucault and Education: Disciplines and Knowledge*. London – New York: Routledge.

Dembowski, J. (1939). Review of the following works by Ludwik Fleck: (1) *Entstehung und Entwicklung einer wissenschaftlichen Tatsache*. Basel (1935); (2) O obserwacji naukowej i postrzeganiu w ogóle. *Przegląd Filozoficzny, 38* (1935); (3) Zagadnienia teorii poznania. *Przegląd Filozoficzny 39* (1936). *Nauka Polska, 24*, 435-439.

Eco, U. (2007). *Sześć przechadzek po lesie fikcji*. (J. Jarniewicz, Trans.). Kraków: Wydawnictwo Znak.

Eribon, D. (2005). *Michel Foucault: Biografia*. (J. Levin, Trans.). Warszawa: Wydawnictwo KR.

Foucault, M. (1999). Sobąpisanie. In M. Foucault, *Powiedziane, napisane: Szaleństwo i literatura* (pp. 303-319). (B. Banasiak et al., Trans.). Warszawa: Fundacja Aletheia.

Foucault, M. (1994). *The Order of Things: An Archeology of Human Sciences*. New York: Vintage Books.

Foucault, M. (1987). Questions of Method: An Interview with Michel Foucault. In K. Baynes, J. Bohman, & T. McCarthy (Eds.), *After Philosophy: End or Transformation?* (pp. 100-118). Cambridge, MA: Massachusetts Institute of Technology Press.

Foucault, M. (1981). Est-il donc important de penser? *La Libération, 30*.

Foucault, M. (1984). Of Other Spaces: Heterotopias. *Architecture, Mouvement, Continuité, 5*, 46-49. Online version of the text: *Of Other Spaces: Heterotopias/Des espaces autres: Hétérotopies*, http://www.foucault.info/documents/heterotopia/foucault.heterotopia.en.html [last accessed: August 12, 2013].

Foucault, M. (1984b). What is an Author? In P. Rabinow (Ed.), *The Foucault Reader* (pp. 101-120). New York: Pantheon Books.

Foucault, M. (1972). *The Archeology of Knowledge and the Discourse on Language.* (M. Sheridan Smith, Trans.). New York: Pantheon Books.

Geertz, C. (1978). Stir Crazy. *New York Review of Books*, January 26, http://www.nybooks.com/articles/archives/1978/jan/26/stir-crazy/ [last accessed: May 17, 2011].

Koczanowicz, L. (1999). Wstęp. In C.C. Lemert & G. Gillan, *Michel Foucault: Teoria i transgresja* (pp. 7-18). (D. Leszczyński & L. Rasiński, Trans.). Warszawa – Wrocław: Wydawnictwo Naukowe PWN.

Lejeune, P. (1989). *On Autobiography.* (P.J. Eakin, Ed.; K. Leary, Trans.). Series: Theory and History of Literature, vol. 52. Minneapolis: University of Minnesota Press.

Lemert, C.C., & Gillan, G. (1982). *Michel Foucault: Social Theory as Transgression.* New York: Columbia University Press.

Martin, R. (1988). Truth, Power, Self: An Interview with Michel Foucault. In L.H. Martin, H. Gutman, & P.H. Hutton (Eds.), *Technologies of the Self: A Seminar with Michel Foucault* (pp. 9-15). Massachusetts: University of Massachusetts Press.

Szacki, J. (2002). *Historia myśli socjologicznej.* Warszawa: Wydawnictwo Naukowe PWN.

Webster's Online Dictionary, http://www.webster-dictionary.org (last accessed: August 8, 2013).

Chapter Six

DO 'PROFESSIONAL' AND 'NON-PROFESSIONAL'
DIMENSIONS OF BIOGRAPHY REALLY EXIST?

DELIBERATIONS BASED UPON *THE AUTOBIOGRAPHICAL NOVEL*
BY MICHAŁ GŁOWIŃSKI
A REFLECTION WRITTEN FOR TWO VOICES

by
Marcin Maria Bogusławski
&
Monika Modrzejewska-Świgulska

Introduction

The issues addressed in this book suggest that we should be dealing with 'professional' and 'non-professional' ways of practicing the humanities, that is—in short—how, for example, a pedagogue or a philosopher go in for humanities and how it is done by e.g. a physicist. The subtitle (clarifying the subject) proposed by the originator of the project, on the other hand, refers mainly to the question of whether the elements of a private biography of a researcher condition his/her professional biography, and if so, in what way.[1] From the outlined alternative, we have chosen the second subject area. The first one is less interesting for us, and apart from that, it would require a detailed empirical research, which could not be effectively carried out in the time frame envisaged for the preparation of the publication. The second subject area seems more interesting as it requires

---

[1]   The Authors refer to the input topic of considerations, namely *Biographies: Between the 'Professional' and 'Non-Professional' Dimensions of the Humanities* (footnote of the editor).

taking a stand on such intriguing matters as the way of understanding the research process in the humanities, or what a biography is, and how it can be comprehended and described.

In our opinion, juxtaposing **the private biography** with **the professional biography** is artificial. A man experiences his or her own existence as **a continuum**, in which the private and the professional sphere are closely related, and often indistinguishable from each other. We mean here the act of 'experiencing,' because in principle, a man does not consciously reflect on his or her existence. The reflection is made only in certain moments—either in **landmark situations** or in situations in which the autobiographical narration serves the **achievement of particular purposes**, for example, applying for a job or a higher degree. Landmark situations result in **comprehensive autobiographical stories**, in which a man makes an auto-interpretation of himself or herself and his/her story. The situations of the second type lead to **fragmentary autobiographical stories** that comprise the elements selected according to the aim which the author wants to achieve, for example life histories or *curriculum vitae*, i.e. something that can be called *a professional biography*, in contrast to *a personal biography*.

From our point of view, the question whether personal and professional biographies condition each other is a consequence of unjustified hypostatizing relating the dimensions, which are instrumentally singled out of the history of a person that forms a uniform whole. If we accepted this juxtaposition, we would, of course, try to show that personal aspects condition professional aspects and *vice versa*. Therefore, the relation would be circular. From the perspective we have decided to choose, the history of a person is a connection of diverse heterogenic elements that make a relational continuum. At the same time, the network of (dynamic, changeable) relations not only binds together the different elements of the history of a person, but it also affects the 'identity' of these elements. Speaking of the 'elements,' we mean, among other things, values, pursued goals, accepted beliefs and norms, remembered experiences and interactions with other people, i.e. the context in which a person is **situated**.

From our perspective, situatedness of a person is relevant and is connected with **the factuality** of his/her life. The relational and interactional character of the life of a person, and his/her historicity, are also essential. Therefore, it is obvious that in the continuum of a person's history, heterogenic elements (that can be placed, for example, in the 'professional' and 'non-professional' sphere) determine one another. The thesis of irreducible narrative character of the (auto)biography (understood as an attempt of a reflective understanding of one's own history, connecting its elements and putting them in order) of a person is also a component of our perspective.

We are going to justify and elaborate on these views analyzing an "autobiographical novel"[2] of Michał Głowiński entitled *Rings of Alienation* (quotes from this book are marked only by a page number). The interpretation will be written for two relatively autonomous voices—one oriented more philosophically, the other—more pedagogically. We hope that, in this way, the topic we have chosen will be clarified in a more thorough way, and also that the complementary character of these otherwise different perspectives will allow to formulate heuristically inspiring conclusions.

1. Głowiński and His Autobiography in the Philosophical Perspective—the Outline of Argumentation[3]

The philosophical tradition that constitutes my frame of reference is socio-hermeneutic ontology. It originates in the hermeneutic philosophy of Martin Heidegger and Hans-George Gadamer, whose assumptions were, however, revised by Barbara Tuchańska and James McGuire (these corrections concern—to put it simply—departing from the individualistic and existential approach of hermeneutic philosophy and moving towards the relational and interactionistic, i.e. the social conception). I am not going to refer directly to Heidegger and Gadamer in my argument, though. A conceptual network established by Jean-Paul Sartre seems to be much more adequate for my purpose. Sartre has a lot in common with the perspective of socio-hermeneutic ontology. From Sartre's standpoint it is possible, for example, to reflect within **the ontic-ontological circle**. A reflection at the ontic level is a reflection on the *status quo*, which is realized namely on what is factually functioning. In my story, it will be a sketch reflection on the fragments of autobiographical findings of Głowiński. A reflection at the ontological level is, in turn, a reflection on the structures which allow for the emergence of what is factual, and thus they are 'conditions of possibility' of what is going on at the ontic level. A presentation (again quite brief) of these structures will be the fundamental aim of my argument. Choosing Sartre to be the patron of the present comments results not only from his close relations to the tradition of hermeneutic philosophy and socio-hermeneutic ontology. There are two more reasons which have influenced such choice of the framework of reference. The first one is that Sartre—as opposed to Heidegger or Gadamer—devoted a large part of his philosophical activity to ontological-ontic biographical research which he called existential psychoanalysis. He elaborated its ontological part

---

2   This term is used as the subtitle of Głowiński's book.

3   This part of the chapter has been prepared by Marcin M. Bogusławski.

mostly in *Being and Nothingness* but also in *Critique of Dialectical Reason*, and he developed its ontic part in many biographies, such as *The Family Idiot*, *Saint Genet*, or, finally, in the autobiography—*The Words*. The second reason is that Głowiński and Sartre belong to the generation joint by the common experience of the war and they operate in a similar political-intellectual context (dominated by Marxism and structuralism). Therefore, it will be interesting to show how—despite their differences—the way of Głowiński's views on history and biography of a person can be expressed in the language of Sartre's ontology.

യ

For Głowiński, the history of a person is not an organized structure, it is rather a chaotic stream of connected elements, which are retrospectively arranged in an ordered structure. Therefore, *The Autobiographical Novel* is "organizing one's life and a sense of self" (Głowiński, 2010, p. 535; cf. also, p. 34), it is creation of (a picture of) one's own identity by comprehensive referring to both introspective experiences and those heard from other people. It is particularly important to base on other people's experience while concentrating on one's own prehistory. "What's heard becomes [then] the subject of personal experience," says Głowiński (p. 5). However, the references to other people's experience as an element of living one's own history accompany a man throughout all stages of his/her life. It is one of the tools enabling to chronologically link all elements considered as fundamental for one's own biography/identity. As an example, it is worth mentioning Głowiński's reference to Marian Skwara's book, thanks to which he could confirm his own memories of Jews' deportation from Pruszków (cf. p. 50), or the reference to the work of Barbara Engelking and Jacek Leonciak, which made it possible for him to determine the time of a violent storm that afflicted Warsaw in 1942. For Głowiński, the storm is one of the best remembered experiences from his childhood, which, in the context of September 1942 (advanced liquidation of the Warsaw Ghetto) and other Głowiński's readings, is regarded as a factor strengthening the memory of how the mechanism that binds the executioner and the victim works (cf. pp. 64-65). In the second case, the reference to the knowledge and experience of other people is used by Głowiński not only to organize his life and himself, but also to make the remembered experiences meaningful. Autobiographical reflection of Głowiński is used to build a linear wholeness from the components creating his history or to make particular elements meaningful, as well as to expose the experience which will be the fundament and mortar for his future fate, in a way the primal experience. Such experience, for Głowiński, is being a Jew, this fact conditioned everything that happened to him and his family. For Głowiński (and his relatives) being a Jew leads to everything—from alienation that influenced his further life to his interests and career (cf. p. 533, 7). On the one hand, the necessity of hiding during the war resulting from being a Jew was the cause of Głowiński's constant solitude. It was also the reason for his discovering the strangeness in relation to other people that posed a threat (for example, a discovery that circumcision constitutes the element identifying him as strange, undesirable). It resulted in permanent neuroticism and considering life as "the highest value, unmatchable" (p. 533). On the other hand, "staying in an apartment in complete isolation from peers, and the lack of what fills the life of a child at this age, somewhat doomed" Głowiński "to this kinds of interest, or even pushed him toward them" (p. 70). Such interests are "intellectual interests" which, in those times, found an outlet in reading books, playing chess or studying a geographical atlas (cf. pp. 70-72). It is worth pointing out that being a Jew is not something Głowiński was just sentenced to. Admittedly, the interiorization of the determinants of 'Jewishness' functioning during the war occurred to a large measure unconsciously in Głowiński's psyche, nevertheless, he constantly was making a conscious decision of accepting or rejecting those elements which were the determinants of his Jewish identity (cf. p. 118). Thus, we witness here a dialectical act of self-selection, projecting one's life on the basis of the category regarded by Głowiński as the fundamental one.

Certainly, it is easy to say that this primal experience influenced Głowiński's career. The question is whether this relation may be presented more precisely. I think it is possible. The relation between extraprofessional experiences and professional interests of Głowiński can be presented at least on two planes.

The first plane concerns the choice of such a method and field of research which allowed him to function relatively autonomously, i.e. on the one hand without exposing himself to fear due to acting inconsistent with the prevailing system, and on the other hand, without the necessity of giving up his own preferences. The method was structuralism, used by Głowiński in his studies on Young Poland's novels or the works of Bolesław Leśmian (Głowiński owes the fascination with Leśmian to historical-biographical circumstances; he discovered his works thanks to a Polish language teacher from an ephemeral, existing only several years, secondary school to which he was sent). One more personal element is connected to structuralism. It is oppositional thinking of himself, characteristic for Głowiński (in his autobiography this is, for example, the opposition of familiarity and strangeness, home and school). This is close to the structuralist methodology, one of the determinants of which is organizing the material with binary oppositions.

The second plane is connected with the unabated feeling of alienation and the reflection on it. This reflection was expressed by Glowiński's studies on the language of the Polish People's Republic which were initially begun for his own pleasure and now are continued during seminars on the analysis of anti-Semitic discourses or in publications about the language characteristic for, *inter alia*, the newspaper *Nasz Dziennik* (*Our Daily Paper*).

An opposite relation is also visible in the autobiographical perspective of Głowiński—i.e. the use of the professional instruments to construct his own history and his own identity (we should remember that such equalization is made by Głowiński himself by identifying the construction of the autobiographical narration with organizing his identity and his life (cf. p. 535)). One of the most characteristic elements visible here is the use of the above-mentioned binary oppositions, and resignation from describing himself in the psychological and idiosyncratic categories, in favor of perceiving himself as *an actant*. Let us note here that 'actant' is a term coined by Algirdas Greimas and it means 'an acting subject,' a figure characterized by specified activities taken up by him/her rather than by his/her inner thoughts (cf. Burzyńska & Markowski, 2007, pp. 291-292). What surprised me during the first reading of *Rings of Alienation* was that the story told by Głowiński-the-author did not give me the insight into the psychological sphere of Głowiński-the-hero. Głowiński-the-author looks at Głowiński-the-hero somewhat from the outside. The story of Głowiński-the-hero is often presented in a cold and neutral manner, the descriptions of emotional states of Głowiński-the-hero are concise and dry. The hero of *Rings of Alienation* is shown as an actant playing the specified roles on the stage of history rather than the subject of his own inner conflicts, experiences, emotions—he plays the role of a Jew battling against the Shoah, a scientist dealing with his own fears and the reality of the time of the Polish People's Republic, a gay dealing with the inner alienation (a description of Głowiński's escape from the entrance of a gay club in the Netherlands) or emerging sexual needs. In this context, it is interesting and worth noting that the reader hardly presumes the emotional needs of Głowiński; it is one more element of the characters' psychology which is completely absent in *Rings of Alienation*. I have to sincerely admit that during the first reading of this book, the 'structuralist' approach of Głowiński to his own history was a bit disappointing to me. Maybe because the reviews of the book emphasized its tragic nature, which I was not able to see enough in the deprived-of-emotions world of Głowiński-the-actant. The first impressions were, however, misleading—the neutrally described world of exclusion is the twice fearsome world. The tragic element is that even in his autobiography, Głowiński protects his strangeness using specified techniques of narration. A close relation between the 'professional' and 'non-professional' components of the history of Głowiński can be noticed here very clearly. In other words, at this point, it can be noted that structuralism was, and still is, a "personal experience" for him (p. 494).[4]

♋

It is time to change the plane of reflections from ontic to ontological. Due to the length limit, also in this case my argument will be just a sketch or a preliminary diagnosis.

Ontological explanation of the ontic characteristics outlined above needs to have a reference to the following categories of Sartre's ontology: *human-reality, being-in-the-situation, project, comprehension, history* and *body*.

Human-reality replaces the category of a subject in my dictionary. The category of a subject is entangled in the context of modern epistemology and, above all, means substantial consciousness opposed to the subject of cognition. The subject understood in the sense presented above is closed in its own inside, i.e.—in short—it is separated "from the flesh, both as the other sphere of human existence (from the body), and from the non-human world (Nature)" (Pobojewska, 2011, p. 12). Talking about the human-reality (the equivalent of *Dasein* in Heidegger's language) allows us to avoid this complication, since this expression leads us simultaneously to individual existence of a person ("being of each one of us is always his/her own being and not being in general" (Folkierska, 2008, p. 143)) and to the fact that an individual always exists in the world, i.e. an individual is an openness inseparably connected with the world. In such perspective, the human-reality cannot be isolated from the network of connections linking it with elements that create the world in which it exists. Therefore, it is impossible to describe the human-reality as an independent, completely sovereign and disembodied subject that follows only the inner dynamics.

I am going to talk mainly about being-in-the-situation instead of being-in-the-world, i.e. in the horizon of "the overall structure of the handy (tool) references" (cf. ibid., p. 143) and the relations to other human-realities (Heidegger's notion of 'being-with'). Because being-in-the-situation is directly connected with the factuality issues (factual situatedness of human-reality), it also makes it possible to talk about various situations,

---

[4] In 2008, Głowiński gave a lecture at the University of Silesia *Strukturalizm jako doświadczenie osobiste (Structuralism as a Personal Experience)*. This issue is also a keyword in his book *Rings of Alienation*. Unfortunately, I did not have a chance to read the text of the paper, however, I think that it could be a confirmation of the thesis of the correlation between the 'professional' and 'non-professional' elements in the history (biography) of Głowiński.

which create either a pluralized image of a world or various worlds (this issue should be examined separately and I am not going to settle it right now). In such perspective, the category 'world' seems to be too universal to me, i.e. too holistic and homogeneous to present the multiplicity of situations and the multiplicity of the worlds connected with it (or the diversity of the world). What is the situation in which the human-reality exists? Following Sartre, I understand 'situation' as the network of relations among variable, heterogenic elements such as climate and earth, race and social class, language, history of the community, habits, whims (cf. Sartre, 2007, p. 586), values, inevitability of death, resistance of things, necessity of work, practical-inertial sphere [*pratico-inerte*] (the matter processed by the human work), social institutions, non-human beings etc. This network constitutes a context which enables the human-reality to self-realize-as-a-project (cf. Cabestan & Tomes, 2001, pp. 56-57). Thus, it is a premise of human-reality freedom.

Being a project means for Sartre "revealing to ourselves, through the possibilities and under some value, of what one is" (Sartre, 2007, p. 693). This "desire of being" (ibid., p. 693) is the aspiration for presenting all the events creating the history of human-reality as a unity. It means that human-reality does not have a given identity, but chooses a sense of who it is (wants to be) and refers everything which constitutes its history to this sense (metaphorically speaking, the projecting sense is a light which allows both to understand the human-reality and to be self-understood by the human-reality). In this sense, the human-reality is absolutely free—its choices are the unfettered self-creation of its own condition. On the other hand, however, the human-reality is never free, contrariwise, it is "enslaved, subjected to necessities, fed to the rights, over which it has no power" and is an easy prey for other human-realities (Kowalska, 2005, p. 91). It just means that the human-reality always self-projects, existing in the situation. Thus, it cannot be who it wants to be—it projects itself while existing in the tangle of relational connections, and this project concerns transforming these connections i.e. transforming the situation, better: co-transforming it along with other self-projecting human-realities. The final effect of these projections depends not only on the human-reality itself, but on all the elements creating the situation in which the human-reality exists. These elements establish an essential context for all actions taken by the human-reality. On the one hand, "man always exists contextually, connected to the outside world through the network of dependencies," and on the other, "man may […] practically process the exterior, and thus realize himself/herself as a spontaneously established project" (Bogusławski, 2008, p. 216) and it helps us to understand the paradox of the 'primal choice,' which—according to Sartre—is the basis of every project. The primal choice is nothing else than a choice of who a person is, the choice of the sense in the light of which a person will interpret himself/herself and his/her history. Such choice is always a sovereign decision. On the other hand, the primal choice is (happens to be?) an interiorization of the features of the human-reality ascribed to it by its companions; at the same time, it is an assimilation of these features and it transforms them "into the abstract form of subjectivity," into the identity considered by the human-reality as its own (Sartre, 2010, p. 58). The further history of the human-reality is, in some sense, an effect of the clash between the dialectics of the sovereign decision gesture and the passive interiorization gesture.

The condition for the existence-in-the-situation and projecting is comprehension. For Sartre—similarly as for Heidegger—comprehension is not a speculative activity or any form of a methodical interpretation, but the way of being, and as such, a fundamentally practical activity (cf. Cabestan & Tomes, 2001, p. 12). Generally speaking, comprehension is a reference to the situations which—following Tuchańska and McGuire—I would like to perceive as "cross-referral to each other of different ways of being" (McGuire & Tuchańska, 1997, p. 157), i.e. as "an aspect of co-being of humans and other beings, connected through relationships and interactions" (Tuchańska, 2007, p. 93). Comprehension understood in this way "is the basis of any human project, and it also makes it possible to grasp the importance of this project." It is so because comprehension, being a specific dialectical movement, enables finding sense of individual actions taken up by a human-reality by pointing to its aim and the initial conditions projecting this aim (Cabestan & Tomes, 2001, p. 11).

Therefore, comprehension "has an ontological source not only in human beings' openness and exposure of the beings, but also in the relational nature of all beings." It is not "the relationship between [...] two opposition beings. It is an aspect of being of (many) entities that interact with one another, their creations and their tools, as well as social institutions making a framework for this interaction, and also that what is the subject of cognition." Comprehension understood as a way of being "obviously has historical character" (Tuchańska, 2007, p. 94). The situatedness of being has similarly historical character. Thus, it is time to ask a question about Sartre's understanding of history. For Sartre, history is not "only a collection of events unfolding in time and constituted by the human past. First of all, it is a process whole-making and unifying, pointing toward the goal (*une fin*) and understanding" (Cabestan & Tomes, 2001, p. 28). The holization process has various dimensions. Sartre elaborates on holization of particular products of the human-reality (e.g. a symphony is holization), on the history as the holization process of human experiences (a project is a process of holization), and on holization as a creation of a situation, i.e. joining various heterogenic elements. In Sartre's opinion, according to the Marxist tradition, the process of holization does not occur on the basis of any inner laws, but it is connected with a human *praxis*, both the individual and the collective one (cf. ibid., p. 58). The source of such understanding of history is, in Sartre's ontology, temporality. There is no room for the detailed lecture on the Sartre's concept of temporality, it is enough to notice that temporality consists of three dimensions—the past, the present and the future. The past is understood by Sartre as an ontological structure which forces the human-reality to be what it is "behind-itself." In other words, the past obliges the human-reality to "follow its being," to catch and project itself on the basis of who/what it was (Sartre, 2007, p. 165). As far as the past means for the human-reality its fulfilled and past identity, the future, for the human-reality, is "what it is to be as something that it cannot be" (cf. ibid., p. 173). Therefore, the future is the plane for the realization of possibilities which could, but do not have to, become the reality.

Very Brief Recapitulation

I think that the transfer of Sartre's category of ontology into the ontic analysis of Głowiński's autobiography is not difficult at all for a thoughtful reader. What is more, I suppose that this transfer is relatively natural, i.e. the act of putting the ontological categories into the ontic analysis does not require any additional 'strained' interpretational procedures. In the perspective presented by me, the human-reality 'exists-in-its-body' in a specified situation, and the situation is a *sine qua non* condition of the understanding of the human-reality and the world by the human-reality itself, including especially the projection of its own sense on the basis of the primal choice. All these elements can be found in Głowiński's autobiography. In principle, he exists in three situations—the war and Shoa, communist Poland and democratic Poland. The war reality—hiding, which was favorable to reading, solitude, traumatization, and, above all, self-defining as a Jew were the primal conditions of the further history of Głowiński, and the starting point of his project. The essential element of the primal choice (project) of Głowiński was also his body: namely circumcision which was considered as an identification factor that determines the form and the course of the process of subjectification. Next, getting aware of the homosexual tendencies strengthened his self-perception as a person opposed to the world, alienated, frightened, worse etc., which was characteristic for Głowiński. Being-in-the-situation was also a starting point for his career, from the choice of the topics of his works (Leśmian, the language of the regime, the language of anti-Semitism), to the choice of the structuralist methodology. In my perspective, the distinction between the private biography of the humanist and his career biography has no sense. The professional activity is one of the elements of the history of the researcher determined (and co-determining!) by the remaining elements of the historical continuum.

2. Głowiński and His Autobiography in the Pedagogical Perspective[5]

My basic reference perspective is pedagogy, social pedagogy, to be more precise (as an academic discipline, not as an everyday educational practice) as well as pedagogy of creativity. In addition, I refer to the fundamental theses of narrative psychology, in order to take a stand on the issues that I have a special interest in, namely, I seek an answer to the question—what a biography is and how it can be understood.

In the first part of my reflections, I will present selected premises of pedagogical disciplines that have been mentioned above, to give a picture—on the example of Michał Głowiński's autobiography—of the perspective which I adopt of understanding the history of a person as a continuum of relationally connected heterogeneous elements, which condition one another. In the second part, I will take a stance on the thesis of the narrative character of an (auto)biography.

The anthropological dimension of social pedagogy is visible in the fact that the center of its interests is **an autonomous individual** that is **independent** but, at the same time, stays in close relation to its **environment** (natural, social, cultural) (see Radlińska, 1961; Cyrański, 1995).[6] Helena Radlińska, the precursor of the discipline, writes about the "diversity of intersecting influences." The process of education is understood as a social phenomenon and as a social activity, and for that reason, answers to questions concerning the factors (belonging to the broadly understood life environment) that influence shaping man's personality are sought. In the traditional concept of social pedagogy, the process of education is described in three basic and interdependent categories of **growth** (natural sphere, somewhat 'spontaneous'), **rooting in** (social sphere, socializing) and **introduction** (cultural sphere, symbolical-communicative). These elements were complemented with the fourth category—**creativity** (sphere of indi-

---

5   This part of the chapter has been prepared by Monika Modrzejewska-Świgulska.

6   Cf. the philosophical categories: 'situation' and 'being-in-the-situation' (the actual location of 'human-reality').

vidual and collective cultural creation) (Piekarski, 2007, pp. 43-57; see also, Szmidt, 2001).

The above-mentioned assumptions result in the necessity (for practitioners as well as academics) of making attempts for the sake of understanding another human being (a ward) and interpreting his/her life with due consideration of the following dimensions: objective (associated with socio-cultural factors) and individual (ontogenetic) (cf. Cyrański, 1995). Therefore, the academic social pedagogy investigates the dependences and interactions between the individual and his/her environment, because the "category of social participation is treated as essential for the understanding of social pedagogy" (Piekarski, 2007, p. 43). Consequently, a social pedagogue will 'observe' how a person reacts to the world, in other words, how he/she co-creates the social and cultural reality through his/her own activity. This interest implies the content of domain problems and, in consequence, the choice of research methods. For example, while analyzing life stories reconstructed on the basis of (auto)biographical materials, either existing (e.g. Głowiński's autobiography) or elicited by a researcher (e.g. narrations from uncategorized interviews), one should, according to the mentioned academic background, treat these narratives as a fragment of a given socio-historical reality that co-determines the shape of peoples' biographies. At the same time, one should take into consideration subjective qualities of the narrators (*inter alia* personal predispositions, experiences, values, undertaken actions, dreams, interests) (cf. Skibińska, 2006; Lalak, 2010).[7]

The discourse of social pedagogy also emphasizes the importance of changeability (situationality) for the course of socialization-educational processes, i.e. the impossibility of separation of the subject from the entire complex context of his/her milieu (Skibińska, 2006; Piekarski, 2007; Marynowicz-Hetka, 2006). This view is followed by other beliefs concerning the education/training process. Not only should it prepare a ward/student to deal with specific situations, but also support their successful development, i.e. shape active (creative) attitude, which is characterized by the independence of actions and thoughts as well as active participation in the dynamic socio-cultural world. Such understanding of education derives from the conviction that human identity is not given but, in a way, constructed, "constantly forming itself" (Walczak, 2010, pp. 255-277). Consequently, the emphasis is put on the processual nature of becoming,

---

[7]   A quotation from Głowiński (2010, p. 25) can serve as an example: "My mother, mentally fragile and prone to isolationism would have possibly become a completely different personality, if she had grown up in more favorable circumstances, not leading to marginalization and the inferiority complex."

which requires the 'teacher—student' relation to inspire the subjects[8] of the educational process to attempt to understand (interpret) themselves and others as individuals being under the same historical influences, and, at the same time, to recognize and respect their individual ways of existence (individuality).

Therefore, choosing social pedagogy as the main conceptual frame of my analysis of selected fragments of Głowiński's *The Autobiographical Novel*, I will try to show the relational and interactionistic nature of the dependencies between the main character of the narrative and different contexts of the story of his life (existential, cultural, social). They might have influenced the process of shaping his attitude towards himself and the world, crystallizing his scientific interests and, as a consequence, his choice of research topics and methodology. Thus, it will be interesting to read *Rings of Alienation* as an attempt to observe his own individuality and uniqueness developing through the interaction with the changing socio-historical reality, which can be defined as the "biographical process of becoming" (Urbaniak-Zając, 2005, pp. 115-127).

For the sake of this analysis, I am also introducing the concept of 'self-creation,' in order to confirm the assumptions mentioned in the introduction. And thus, I refer to the selected premises of pedagogy of creativity, which are connected with the theories of creative (active) attitudes, formulated mainly by social pedagogues, humanistic psychologists and psychologists of abilities (cf. the above reference to the concept of creative, active attitude) (Szmidt, 2001). At present, justifications of the theories of creative attitudes are sought in the research of positive psychology, in which the following psychological attributes: ability to forgive, gratitude, spirituality, wisdom, hope, sense of purpose, humility, self-esteem and creativity, are treated as cardinal virtues that facilitate the feeling of happiness, autonomy and fulfillment (see Trzebińska, 2008; Modrzejewska & Szmidt, 2013). By contrast, according to the social pedagogy tradition, human resources are defined as individual and social forces that make individual and collective activities dynamic. The term 'self-creation' means intentional and deliberate activities undertaken in order to shape one's own identity (self-formation), i.e. constructing a coherent image of oneself, especially through organizing biographical experiences and adding meanings to them. It becomes possible due to the insight into one's own needs, motivations and emotions, that is self-reflection and self-cognition.

---

[8] In pedagogical tradition, the term 'subject' is used, *inter alia*, to describe a non-subjectified way of existence in educational relations and "the ordinary meaning to determine man as the subject of action, such as education, learning, etc." (Męczkowska-Christiansen, 2006, p. 15).

I assume that self-creation processes can enable projecting[9] one's life through the reflection in the past and present, and anticipation of the future, i.e. through constant addressing the issue—'who am I in the changing and dynamic world?' The process of developing one's own identity is, in other words, an attempt at self-understanding, mainly through the confrontation and connection with other people and the milieu (cf. Schulz, 1900; Pietrasiński, 1990; Uszyńska-Jarmoc, 2007). I believe that the term 'self-creation' should be, first and foremost, associated with positive ways of realizing one's biography, i.e. creative transformation of the found reality with the use of one's potential (resources, strengths), so as to lead a life more satisfying in its various dimensions (cf. Smolińska-Theiss & Theiss, 2010). To put it differently, the uniqueness and fruitfulness of the life of a person depends not only on the socio-historical context and structural features of his/her personality, but also, to a great extent, on the decisions taken and their realization through (extra)ordinary actions (*praxis*).

଼৪

Through the reference to Głowiński's (narrator) autobiography, I will attempt to increase the plausibility of the adopted assumptions concerning the history of a person and support the pedagogical dimension of this interpretation. Generally speaking, the pedagogical perspective is expressed in the following questions: in what way does the narrator 'learn the life,' or how does he 'deal' with the unpredictability of his fortune? How does he react to the world? And finally—in what way does a given social order shape his identity and individuality? Therefore, I will address the issue of **historic and social contexts of the learning processes** in an 'autobiographical novel' of the abovementioned author.

The narrator had come to live and grow up in four different **socio-historical contexts**, which forced upon him the necessity of adjusting to different living conditions—Poland just before World War II, wartime, communistic Poland and, later on, democratic Poland. For this reason, Głowiński, reconstructing his own fate, refers to the specific socio-cultural contexts that he has witnessed and experienced. Thanks to that, his autobiographical recollections can be read as a story of the Holocaust, times of the Polish People's Republic (PRL) or the birth of the Polish democracy, as well as his interpretation of politics, culture or the society of these times (cf. Głowiński, 2010, p. 88, 95). His own experiences give the narrator the right to comment on the public affairs, especially on difficult Polish-Jewish relations, which he states straightforwardly: "[...] based on my

---

[9]   Cf. earlier philosophical findings.

own experience, I have the right to write about this kind of terrible approach to Jewish matters during the occupation" (p. 89).

Treating an autobiography as memories of the person who, throughout the whole life, has been accompanied by feelings of alienation and otherness is much more interesting for a pedagogue, though. This acknowledged scientist and writer felt alienated in Poland as a Jew, especially during the war. He also experienced lack of understanding and separation as an atheist, homosexual, member of the Jewish community and as a person suffering from claustrophobia, which hindered him from, among other things, the participation in mass meetings and strikes against the communistic regime, and limited his integration with the striking academic community. The socio-historical context and Głowiński's subjective conditioning caused his inability to be fully himself in relations with others. He was forced to live in constant hiding, wearing 'masks,' experiencing loneliness, hence the title 'rings of alienation'—anti-Semitism, homosexuality, claustrophobia. The fundamental experience, constitutive for his existence in consecutive 'rings of alienation,' is the narrator's Jewish descent, which among other things, determined his being in permanent danger during the war, and the necessity to hide both physically and symbolically (concealing his Jewish identity). The narrator describes that experiences in the following way: "[...] this situation triggered in me the attitude that I refer to as the snail syndrome. At any time, I wanted to stick to the belief that I can close myself inside the shell and not expose out anything that belongs to me. This could, if not eliminate, at least mitigate the sense of danger" (p. 79).

Since the Holocaust, Głowiński had to get by with a secret, learnt how to live with it, and this gave rise to the awareness of alienation, strengthened by further life experiences: times of Stalinism (anti-Semitic propaganda), discovering his homosexual inclinations, claustrophobia. He is still aware that the war, occupation, the Holocaust did not "end in him," what is more, they have "lasted for many years, taking a variety of shapes, they are lasting and will continue to last until my last day" (pp. 106-107). This constant necessity of hiding the true 'I', especially in the periods of childhood and adolescence, shaped the narrator's identity, the identity of a 'snail,' and influenced his further life decisions that were the continuation of the war experiences. One example of such decisions could be the choice of Polish studies and his profession that favored working in solitude, with no need for "coming out of one's shell." Although he admits that even in his childhood he was displaying introvert tendencies, they were still strengthened during the war: "In childhood and youth, I did not like and I could not be among my peers, I felt bad among them, bothered from the beginning by alienation" (p. 55). The narrator copes with alienation, fear or the feeling of otherness, by going in for activities associated with areas that, first of all, allow him to function independently, and, secondly, are ruled by relatively clear principles, help him to discover the hidden meanings and, what is more, favor the creative transformation of those meanings. The "treasures" that helped him to survive "the vegetation of unusual daily life" of the war, include: a chess set, a German atlas of the world and Grimms' fairy tales, and after the war: reading encyclopedias and books "about dying and suffering," and listening to music, which was an antidote to the feelings of "futility" and "apathy." Later on, he was mainly engaged in literary theory, journalism as well as writing reminiscent prose and autobiography. Reminiscent prose may function as a self-therapy, it brings relief and helps to understand one's history, which cannot be acquired through psychological therapy, however, I think that it is due to this therapy that the writer 'opens' to the therapeutic qualities of literature.

The wartime experience, in Głowiński's biography, certainly includes 'nuclear episodes' i.e. crucial events that substantially influence the transformational processes and life history. In autobiographical narrations, these are usually some turning points which very often play the main part in a life story (Tokarska, 1999). Głowiński writes, "the experiences of those times will continue to come back to this story, it is impossible, for various reasons, not to refer to them" (p. 107).

In Głowiński's narration, **the social context** of learning processes is obviously connected with the socialization-educational processes, i.e. with the functioning within social groups, such as: family, school, orphanage run by nuns during the war, college, career and, to a lesser extent, with the presence of other people.

Głowiński is aware that, as Peter L. Berger and Thomas Luckmann (1967, p. 129) notice, "The individual [...] is not born a member of society. He is born with a predisposition toward sociality, and he becomes a member of society," and the personality- and identity-creating processes are closely linked to being assigned to a specific place in the world. At the beginning of the tale of his fate, the author of *Rings of Alienation* roots it in the history of the members of his family: "When we speak of short prehistory (about the social and professional origin [of ancestors—M.M.-T.]), social history becomes particularly important, because knowing to which social, national, local group the ancestors belonged, one can indirectly learn this and that about them" (p. 6). "How [...] could I describe my family in the most general outline? First, I would say that it is Jewish—this is obviously the most significant feature, because it determines the rest and it defines my family at the highest level" (p. 7), he adds.

People that could be perceived as important for Głowiński's biography, practically do not appear. Only his parents, in other words the 'significant others,' are portrayed there in a more detailed way. During the primary socialization, the social world reaches the individual through the parents, i.e. the first world of an individual is created. "To disintegrate the massive reality internalized in early childhood" of this world, "severe biographical shocks" are needed (Berger & Luckmann, 1967, p. 142), which the narrator (Głowiński) attempts to accurately reconstruct in the chapters devoted to his war memories.

Coming back to the issue of the presence of people important for the writer's biography, it is necessary to mention that the descriptions of teachers are laconic, except for the one concerning the priest who had strongly anti-Semitic views. The appearance of these people is sanctioned rather by the chronology of Głowiński's life story than by their significance for his choices. The influence of these people can be defined as incidental, except for the Polish teacher from high school. Reading the autobiography, one has the impression that the independent choices, next to the socio-historical background, are the main catalysts for decision-making; and so we learn about the agent of the action (*actant*) and not about the psychological individual. We have no insight into the psychological sphere of the main character of the narration, into his emotions and inner conflicts, as their descriptions are terse and unemotional. This can be connected with the linguistic research of Głowiński-structuralist, because, as Roland Barthes (1975, p. 256) states, "From the very first, structural analysis showed the utmost reluctance to treat the character as an essence, even for classification purposes." It appears that the author wants to recreate the 'syntax' of his behaviors presented in the autobiography, to draw the path of his choices. Therefore, we observe him as a person functioning in specific social contexts, realizing subsequent social roles—of a child raised in a Jewish family, and further, of a linguist, writer, publicist and gay, hiding due to his descent, because "singleness of life line is in sharp contrast to the multiplicity of selves one finds in the individual in looking at him from the perspective of social role" (Goffman, 1986, p. 63). At the same time, we see him as a lonely individual, aware of his stigma, and, as a result, having difficulties with establishing close relations with others, and protecting the hidden, sometimes embarrassing information on him from being used by wrong persons. The minimization of the psychological sphere in the biography seems obvious. However, the author can return to traumatic events, which are still 'alive' in his memory, and describe what is 'indescribable.' Distancing himself, he speaks of the world which distorted human fate. In this way, the reader is, in a sense, given the 'freedom' of interpretation, undisturbed by excessive sentimentalism and emotionality, so that he/she can adopt a specific attitude towards the depicted world, and especially towards the topics that are difficult to express (associated with Polish-Jewish history and narrator's homosexuality).

To sum up, one may venture a pedagogical conclusion that Głowiński's *The Autobiographical Novel* depicts the process of self-education, i.e. gaining the abilities of self-understanding and self-realization. Therefore, through an autobiography we learn about the main character, who is involved in historical events and whose individual growth evolves from the involuntary adjustment to the existing circumstances (e.g. conditions during the war, which actually took away the possibility of successful development and forced Głowiński to exist in a new, terrifying reality), through internalization of social norms and cultural values, towards a more and more reflective and conscious participation in his own life and the socio-cultural world, and, as a consequence, creative transformation of both spheres due to his own activities (cf. Piekarski, 2007), because "man earns the personality with his own effort, thanks to his creative attitude" (Radlińska, 1947, p. 21).

For pedagogues, the analyzed autobiography may be interesting, as we 'meet' the narration's main character, who experiences difficulties and suffering, adjusts to new living conditions and sets goals (cf. Urbaniak-Zając, 2005). He takes the reader through the process of becoming a mature and self-aware person, the process depending on, among other things, overcoming life perturbations and alienation, which leads the main character from existing in the title 'rings of alienation,' to active and open communicating of himself to others. It can be observed on the two layers of the novel, intertwining each other throughout the whole narration: **reporting reality** (through the description of episodes, experiences, people, choices and social relations, which make up Głowiński's life story and are clearly visible in the activities taken up for the practiced scientific field, cooperation with the academic community, running a PhD seminar, establishing friendly or partnership relations) and also **constructing his own image for the purpose of the story**, i.e. in descriptions of painful war experiences, views expressed on historical and social topics, commentaries on certain people's behaviors, recalling preferences of cultural texts or revealing his homosexuality ("I have decided to unveil this realm of my life for this reason only that I care for this story of myself to be free from gaps and falsehoods, from inauthenticity and pretending" (p. 147)).

CB

It is time to show more precise relations between *the experiences of an individual* and *the sphere of professional activity*, in accordance with the view that "science is not practiced by beings abstracted from the world, but living persona—people situated culturally and socially" (see Kafar, *Around 'Biographical Perspectives'* in this book).

Głowiński lives in a permanent sense of danger and fear that others may discover his well-kept 'secrets'—Jewish descent, homosexual tendencies, claustrophobia. He makes choices connected with both the sphere of 'non-professional' and 'professional' activities, which not only secure his quite independent, inconspicuous existence, but also give him a possibility of realizing his own cognitive interests. The relation between 'non-professional' and 'professional' experiences is visible in three spheres: the choice of studies, and thus the field of interests, and then, within its framework, in the choice of specified topics and research methodology—*inter alia* structuralism and narratology. A question appears: 'To what extent could the knowledge of structuralism have influenced Głowiński's own history reconstruction?' or, in other words, 'Which elements of the autobiography are connected with structuralism or narratology?' It is essential to remember that "each creator is a specific personality in terms of both the psychological and sociological dimension, he bears the individual predisposition to shaping a language in a particular way" (Głowiński, Okopień-Sławińska & Sławiński, 1986, p. 142). Admittedly, Głowiński clearly emphasizes that the cognitive goals have always been more important to him than the uncritical and orthodox use of the specified theoretical literary doctrines (he writes in his autobiography: "I did not specify that [...] I did not think of myself (and I do not think!) in this way, all self-characteristics of this type seemed to me, and still seem, funny indeed" (p. 242)), however, it is worth to consider whether, in *The Autobiographical Novel*, it is possible to nail a workshop of a structuralist and a 'narratologist,' which may concern, *inter alia*, the category of "a virtual receiver,"[10] oppositional thinking,[11] resignation from the psychological description of himself or a linear construction of the time layout in the text (cf. the fragment prepared by M.M. Bogusławski; see also, Głowiński, 2004).

Polish structuralists, including Głowiński, have been interested in the social functioning of a literary work, so the theories of literary communication have been the most important in their reflections. Thus, they have

---

[10] See Głowiński (1977).

[11] Apart from the already noticed oppositions, we can point to those accompanying the ghetto descriptions: 'crowd—space,' 'heat—cold,' 'dirt—cleanness' (the description of German soldiers and ghetto dwellers) and the construction of the autobiography itself in which we can distinguish the times before and after war.

intended to conduct "precise examination of the set of relations inscribed in the literary message and functioning between the sender and the recipient" (Burzyńska & Markowski, 2009, p. 293). There have been attempts to define the general conditions of the possibility of literary communication e.g. the author's comments, a theoretical figure of the recipient written into the literary piece ("the assumed recipient"). We read in *Rings of Alienation*: "[...] the issue of the recipient of a literary piece as an element of its structure was dealt by me a little later, in the second half of the sixties" (p. 243). Głowiński on a number of occasions emphasizes the purposefulness of the used composition of the presented events, for example: "I return to taking about [...]," "I am writing about this to emphasize that [...]," „before I talk about [...], I would like to devote some place to something else [...]," "I wrote about it so extensively [...], because [...]," „as I have mentioned above [...]." Thanks to such means, the reader has an impression of taking part *in statu nascendi* in the process of creating the autobiography. Therefore, one can say that the author wanted the reader to understand his intentions and the interpretation of his life history, and to properly interpret his scale of values and the hierarchy of importance. The above-presented 'devices' connected with the communicative situation of the literary piece and the discussion of difficult topics (such as 'the nightmare' of the ghetto, discovering the sexual identity, acute claustrophobia, fear concerning his own fate in the Polish People's Republic) assume that the work is addressed to a sensitive recipient, who does not generalize and is open for an individualized message. The author resigns from colorful stylistics in favor of a specific, referring language. He cares for word precision, to be well understood. The subsequent chapters of the autobiography are not only built according to the rule of a simple chronology in life, but also reveal new dependencies between various elements of the author's life history.

The extraprofessional experiences of Głowiński might have influenced the choice of a field and methodology; they also might have had an impact on the subject matter in his research works. The researcher has written essays on the language of the Polish People's Republic (e.g. *Mowa w stanie oblężenia* (*Speech Under Siege*)) and was the originator of seminars and articles demystifying anti-Semitic discourses. The issue of war has been an object of Głowiński's interest since his childhood, and, as an adult, he has chosen to make it one of the topics of his reminiscent prose (e.g. in *Czarne sezony* (*The Black Seasons*)).

On the one hand, all the things we experience influence our professional choices, including the choice of the field of science, the issues and methodology, on the other hand, at the same time, they become a continuum of extraprofessional experience. In case of Głowiński-structuralist, it influenced the way of the autobiographical experience reconstruction.

CB

Below, I will refer to the thesis concerning the narrative character of a(n) (auto)biography of a person, perceived as an attempt of reflexive understanding and ordering one's own life history by linking its elements since, as Erving Goffmann (1986, p. 62) reminds, "biographies are very subject to retrospective construction."

The title of Głowiński's autobiography informs us that it will be a story about life. What could it mean? **Narrative psychology** assumes that "life is a story" ("it is structured like a story"), and "if life is a story, it expresses the personal interpretation of events, and also the motives, feelings and aspirations" (Oleś, 2008, p. 37; cf. also, Trzebiński, 2002). Reading *Rings of Alienation*, we find out that the author wanted to "control" his life rather than "present it" ("reflect" it); he wished to show both "the main routes" and go deep into the "nooks and crannies" of his life, trying to, in this way understand and order his own history ("I know that an autobiography is a result of a compromise between what is important and what is remembered" (pp. 532-533)). Głowiński orders his life according to his own key, he selects pieces of information which are important to him, and therefore, he gives the chosen events and experiences a subjective meaning. Thus, the autobiography is a narration, which pours the sense to life, explains it subjectively but, at the same time, creatively. It allows the writer to construct his own narrative subjectivity and identity. In Głowiński's novel, there are categories ordering his biographical experience; they are described in a more detailed way by Jacek Piekarski (2007): **"continuity"** (linking personal past and present and, on the basis of this, projecting the future, constructing a life history which guarantees the sense of continuity and coherence, merging various experiences occurring at the same time); **"complexity"** (visualization and summoning experiences in various social groups); **"originality"** (noticing his own individuality in decision making, especially in the context of a social position); **"realization"** (creating his own portrait as a competent person dealing with difficult situations, realizing his own goals and values) (see also, Oleś, 2009).

I understand narrative construction of an identity as self-creation activities bringing new and positive images of one's own life, i.e. a reinterpretation of one's own life history, which helps to understand and tame the aspects that are painful, accidental and remote in time. In Głowiński's narration, self-creation involves referring to personal experiences as the source of the knowledge of oneself, the world around, and the relations between these two entities (cf. Dziemianowicz & Kurantowicz, 2005). The author experiences suffering, but 'turns' it to something beneficial to him, i.e. he uses it to develop. He answers his own questions: 'Who am I?' and 'What can I connect with the general goal of describing my life?' It appears that, because of the above, *The Autobiographical Novel* is an example of "successful biographization," "that allows to protect a positive self-image" (Urbaniak-Zając, 2005, p. 122). In narration studies, it is an example of a pedagogical perspective which analyses the possibility of an intentional creation of one's own biography (of life) (cf. Urbaniak-Zając, 2005; Lalak, 2010).

ଔ

To sum up, it can be stated that referring to the categories of social peda- gogy and creativity pedagogy for the purpose of analyzing Głowiński's au- tobiography is rather obvious, and—I hope—convincingly outlined. From the perspective of the mentioned pedagogical disciplines, the subject exists in a specified socio-historical context (situation). To be able to understand another person (also through the texts written by him/her), especially ac- cording to the principles of hermeneutically oriented social pedagogy, it is essential to ask about a historical location of the subject and individual- ized dimension of his/her existence (see Cyrański, 1995; Walczak, 2010). In my perspective, a human being, despite his/her strict dependence on the environment, time, and influences of social groups, activates his/her development potential (individual strengths), thanks to his/her own cre- ative actions (active attitude, independence, self-reflection) he/she becomes more and more autonomic, i.e. creates his/her own life (self-creation), gives it a subjective meaning, communicates it to the others in order to, among other things, make these meanings common and build fruitful interpersonal relations, and, what is more, he/she takes actions which go beyond the indi- vidual goals (cf. Radlińska, 1961; Szmidt, 2001; Piekarski, 2007).

Conclusion

In our narration, we concentrated on the question what a biography is and how it could be understood. We were especially interested in the issue of its 'professional' and 'non-professional' dimension. We do not agree to artifi- cial contrasting *a personal biography* with *a professional biography*, so we tried to show, each of in line with his/her own traditions and the perspective of his/her own field of studies, that a human being experiences his/her life as **a specific continuum**, in which *the personal* and *professional* elements are often impossible to be distinguished, because we experience heterogenic elements that build our life in a simultaneous manner. The professional issues do not exist only when we enter a laboratory, a school class, or an office. A strict division into the personal and professional biography may be questionable, because it is based on the assumption that a person is someone else in his/her personal and professional life. We distinguish these two spheres only thanks to self-reflection, writing autobiographies or telling other people about our life, so as to order and understand things we have experienced.

At the end of our considerations, we present a short summary of the ideas discussed above, in the form of the most important theses:

- the proposed perspectives—philosophical and pedagogical—perceive a person as a human-reality/acting subject (*actant*) introducing conscious changes, being in relations with others, communicating;
- to understand the uniqueness of a life of a person, one should not only concentrate on the inner features and mental processes dynamics of the subject in the analyses of a(n) (auto)biography, but should treat a person as a whole, i.e. as a 'traveler,' existing in a specified socio-cultural context (network), interacting with other 'travelers' (being-in-the-situation/an individual in his/her milieu);
- we do not equate an (auto)biography with the subsequence of events or a linear causality only, but we perceive it correlatively, as a narration, i.e. a reflective understanding of one's own history;
- we understand the identity/personality in dynamic, processual, relational and linguistic dimensions;
- the identity is constructed (chosen), thus we can realize ourselves-as-the-project (to be a project)/take auto-creative actions, but always in a specified situation, which both limits and allows self-projecting/a creative (active) attitude; the experiencing human-reality/subject shapes and is shaped simultaneously, refers to himself/herself and to others;
- the understanding (referring) to ourselves and to others is a lifestyle/self-realization which is a practical activity and depends on what is happening to our body;
- philosophers and pedagogues "very often talk about the same human issues," although they speak different languages (cf. Jastrzębski, 2005).

ଔ

Acknowledgments: We would like to thank Jacek Świgulski, who actively took part in our discussions, thus bringing a new perspective to our reflections.

References for the fragment prepared by Marcin M. Bogusławski

Bogusławski, M.M. (2008). "Podmiot" u Sartre'a i Canguilhema: Notatki. *Forum Oświatowe*, Special Volume, 213-223.

Cabestan, P., & Tomes, A. (2010). *Le vocabulaire de Sartre*. Paris: Ellipses.

Folkierska, A. (2008). Pytanie o podmiot: Między Heideggerem i Hessenem. *Forum Oświatowe*, Special Volume, 141-152.

Kowalska, M. (2005). *Demokracja w kole krytyki*. Białystok: Uniwersytet w Białymstoku, Wydział Historyczno-Socjologiczny.

McGuire, J.E., & Tuchańska, B. (1997). Sytuacja poznawcza: Analiza ontologiczna. In J. Goćkowski & M. Sikora (Eds.), *Porozumiewanie się i współpraca uczonych* (pp. 141-159). Kraków: Secesja.

Pobojewska, A. (2011). Wprowadzenie. In A. Pobojewska (Ed.), *Współczesne refleksje wokół kartezjańskiej wizji podmiotu* (pp. 5-32). Łódź: Wydawnictwo Akademii Humanistyczno-Ekonomicznej.

Sartre, J.-P. (2010). *Święty Genet: Aktor i męczennik*. (K. Jarosz, Trans.). Gdańsk: Słowo/Obraz Terytoria.

Sartre, J.-P. (2007) *Byt i nicość: Zarys ontologii fenomenologicznej*. (P. Mróz et al., Trans.). Kraków: Wydawnictwo Zielona Sowa.

Tuchańska, B. (2007). Socjo-historyczna hermeneutyka poznania zamiast epistemologii: Projekt badawczy i edukacyjny. In M. Hetmański (Ed.), *Epistemologia współcześnie* (pp. 85-100). Kraków: Towarzystwo Autorów i Wydawców Prac Naukowych Universitas.

References for the fragment prepared by Monika Modrzejewska-Świgulska

Barthes, R. (1975). An Introduction to the Structural Analysis of Narrative. *New Literary History*, 6, 2, 237-272.

Berger, P.L., & Luckmann, T. (1967). *The Social Construction of Reality: A Treatise in the Sociology of Knowledge*. New York: Anchor.

Cyrański, B. (1995). *Rekonstrukcja aksjologicznych podstaw pedagogiki społecznej z zastosowaniem metody interpretacji hermeneutycznej*. Doctoral dissertation, Faculty of Educational Sciences, Univeristy of Łódź.

Dziemianowicz, M., & Kurantowicz, E. (2005). Filozoficzne inspiracje dla pedagogicznych badań biograficznych (dwugłos badaczy). In P. Dehnel: & P. Gutowski

(Eds.), *Filozofia a pedagogika: Studia i szkice* (pp. 107-128). Wrocław: Wydawnictwo Naukowe Dolnośląskiej Szkoły Wyższej.

Głowiński, M. (2004). Wokół narratologii. In M. Głowiński (Ed.), *Narratologia* (pp. 5-12). Gdańsk: Wydawnictwo Słowo/Obraz Terytoria.

Głowiński, M. (1977). *Style odbioru: Szkice o komunikacji literackiej.* Kraków: Wydawnictwo Literackie.

Głowiński, M., Okopień-Sławińska, A., & Sławiński, J. (1986). *Zarys teorii literatury.* Warszawa: Wydawnictwa Szkolne i Pedagogiczne.

Goffman, E. (1986). *Stigma: Notes on the Management of Spoiled Identity.* New York: Simon & Schuster.

Jastrzębski, J. (2005). A może by tak przestać zadawać pytania? (o filozofię i pedagogikę). In P. Dehnel & P. Gutowki (Eds.), *Filozofia a pedagogika: Studia i szkice* (pp. 28-41). Wrocław: Wydawnictwo Naukowe Dolnośląskiej Szkoły Wyższej.

Lalak, D. (2010). *Życie jako biografia: Podejście biograficzne w perspektywie pedagogicznej.* Warszawa: Wydawnictwo Akademickie Żak.

Męczkowska-Christiansen, A. (2006). *Podmiot i pedagogika: Od oświeceniowej utopii ku popkrytycznej dekonstrukcji.* Wrocław: Wydawnictwo Naukowe Dolnośląskiej Szkoły Wyższej.

Modrzejewska, M., & Szmidt, K.J. (Eds.) (2013). *Zasoby twórcze człowieka: Wprowadzenie do pedagogiki pozytywnej.* Łódź: Wydawnictwo Uniwersytetu Łódzkiego.

Marynowicz-Hetka, E. (2006). *Pedagogika społeczna.* Warszawa: Wydawnictwo Naukowe PWN.

Oleś, P.K. (2009). *Wprowadzenie do psychologii osobowości.* Warszawa: Wydawnictwo Naukowe Scholar.

Oleś, P.K. (2008). *Autonarracyjna aktywność człowieka.* In. J. Bernadetta, K. Gdowska, & B. de Barbaro (Eds.), *Narracja: Teoria i praktyka* (pp. 37-52). Kraków: Wydawnictwo Uniwersytetu Jagiellońskiego.

Piekarski, J. (2007). *U podstaw pedagogiki społecznej: Zagadnienia teoretyczno-metodologiczne.* Łódź: Wydawnictwo Uniwersytetu Łódzkiego.

Pietrasiński, Z. (1990). *Rozwój człowieka dorosłego.* Warszawa: Wiedza Powszechna.

Radlińska, H. (1961). *Pedagogika społeczna.* Wrocław: Zakład Narodowy im. Ossolińskich.

Radlińska, H. (1947). *Oświata dorosłych: Zagadnienia, dzieje, formy, organizacja.* Warszawa: Ludowy Instytut Oświaty i Kultury.

Schulz, R. (1990). *Twórczość: Społeczne aspekty zjawiska*. Warszawa: Wydawnictwo Naukowe PWN.

Skibińska, E.M. (2006). *Mikroświaty kobiet: Relacje autobiograficzne*. Warszawa: Uniwersytet Warszawski, Wydział Pedagogiczny.

Smolińska-Theiss, B., & Theiss, W. (2010). Badania jakościowe—przewodnik po labiryncie. In S. Palka (Ed.), *Podstawy metodologii badań w pedagogice* (pp. 79-102). Gdańsk: Gdańskie Wydawnictwo Psychologiczne.

Szmidt, K.J. (2001). *Twórczość i pomoc w tworzeniu w perspektywie pedagogiki społecznej*. Łódź: Wydawnictwo Uniwersytetu Łódzkiego.

Tokarska, U. (1999). W poszukiwaniu jedności i celu—wybrane techniki narracyjne. In A. Gałdowa (Ed.), *Wybrane zagadnienia z psychologii osobowości* (pp. 169-204). Kraków: Wydawnictwo Uniwersytetu Jagiellońskiego.

Trzebińska, E. (2008). *Psychologia pozytywna*. Warszawa: Wydawnictwa Akademickie i Profesjonalne.

Trzebiński, J. (2002). Narracyjne konstruowanie rzeczywistości. In J. Trzebiński (Ed.), *Narracja jako sposób rozumienia świata* (pp. 17-42). Gdańsk: Gdańskie Wydawnictwo Psychologiczne.

Urbaniak-Zając, D. (2006). W poszukiwaniu kryteriów oceny badań jakościowych. In D. Kubinowski & M. Nowak (Eds.), *Metodologia pedagogiki zorientowanej humanistycznie* (pp. 209-222). Kraków: Oficyna Wydawnicza Impuls.

Uszyńska-Jarmoc, J. (2007). *Od twórczości potencjalnej do autokreacji w szkole*. Białystok: Wydawnictwo Uniwersyteckie Trans Humana.

Walczak, A. (2010). Tożsamość narracyjna i 'choroby serca' nauczyciela. In J.M. Michalak (Ed.), *Etyka i profesjonalizm w zawodzie nauczyciela* (pp. 253-280). Łódź: Wydawnictwo Uniwersytetu Łódzkiego.

Common references

Głowiński, M. (2010). *Kręgi obcości: Opowieść autobiograficzna*. Kraków: Wydawnictwo Literackie.

Burzyńska, A., & Markowski, M.P. (2007). *Teorie literatury XX wieku: Podręcznik*. Kraków: Wydawnictwo Znak.

Chapter Seven

ETHNOGRAPHIC EXPERIENCE AND THE POLITICS OF SITUATEDNESS

by Marta Songin-Mokrzan

Over the last few decades there have been intensive ongoing discussions in anthropology on the limits of cognition, the role and function of the subject in the production of knowledge, as well as the representations of the cultural reality in ethnographic texts. Within the research interests there have been included issues related to the status of ethnographic truth and knowledge acquired within the discipline. The concept of an autonomous and rational subject has been challenged, the possibility of objective description of reality has been questioned, and the correspondence theory of truth has been rejected together with the belief in the universal nature of human reason (Buchowski & Kempny, 1999, p. 11). The content shape of these debates has been influenced by the criticism of the interdependencies existing between anthropology, colonialism and imperialism (the so-called political crisis), feminist research (taking up such issues as the impact of gender and other 'markers of identity' on the cognitive process) and the discovery of "literary mechanisms of producing the presented worlds" (Brocki, 2008, p. 9) closely related to the epistemological crisis. These considerations have largely changed the image and the perception of anthropology; they have become a starting point for the reflection in the field of the theory, methodology and methods of ethnographic research. This has led to the emergence of a number of trends of an interdisciplinary nature and the creation of various forms of ethnographic writing and innovative research techniques. In addition, gradual moving away from the positivist to the hermeneutic or phenomenological research model has become apparent, as well as turning to the critical theory.

With the discovery of the dialogic nature of knowledge, emerging in the process of negotiations of meanings and senses, the researchers have made the subject of the overview the anthropology itself, which, in their opinions, produces Otherness through locating it in the Western cultural discourses. Moreover, within these considerations, the Other is also the "anthropologist, who is his/her own anthropological imagination" (Kaniowska, 1999, p. 135). Therefore, it can be seen that a unique role in the contemporary anthropology is played by reflexivity, understood as a kind of 'self-deconstruction' of the discipline combined with the development of its self-awareness. The assertion that reality is interpreted by culturally situated subjects allows also to see reflexivity as a "critical scrutiny of the self" (Okely, 1992, p. 2), which in turn provokes reflection on the role of biographical elements in the production of knowledge.

In this chapter, I will consider what methodological effects are brought about by the changes described above, and how they contribute not so much to a fuller understanding of what (anthropological) knowledge is and how it is created, but to a better, more in-depth comprehension of the reality studied by anthropologists. This matter I will examine in relation to the fundamental anthropological category, namely **experience**. I will be interested on the one hand in the question of relations between *individual experience of a researcher* and *ethnographic experience* ('professional' one), and on the other hand, in the issue of *political conditions of anthropological knowledge*.

♃

As noted by Wojciech J. Burszta (1992, p. 141), "since the time of [Bronisław] Malinowski, the method of the so-called participant observation has been meant to establish a delicate balance between subjectivity and objectivity of the knowledge possessed by an anthropologist. Personal experience of a researcher and particularly participation and empathy for the natives were considered central to the process of understanding the society and culture in question. They were, however, simultaneously limited by the impersonal standards of observation and 'objective distance.'" It can be clearly seen that in this fieldwork method constitutive for anthropology, namely participant observation, there lies a certain paradox. The researcher is required both to perform the total 'immersion' in the indigenous culture and keep the critical distance towards it. Therefore, since the dissemination and institutionalization of the long-term field research in the humanities, a boundary difficult to define between 'personal' and 'professional' has manifested itself. This division is reflected in the kinds of elaborated texts: studies of communities and—written somewhat at the margins of the mainstream—'memories' of scholars (diaries, dramatized stories, poems, novels, etc.). Interestingly, in a large part because of Bronisław Malinowski and his diaries first published in 1967 (cf. Malinowski, 1967), the category of ethnographic experience is subjected to further reconceptualization. The stream of personal experiences of a researcher, previously hidden under the mask of objective description, becomes the leitmotif of the confessional and impressionist experimental ethnographies (cf. Van Maanen, 1988, pp. 73-100).

According to Katarzyna Kaniowska, a change in the understanding of the status of experience in anthropology goes hand in hand with an increase in the role of hermeneutics and postmodernism on the grounds of the discipline. Kaniowska (2006, p. 20) writes, "From trusting the perception through the senses we have moved to understanding experience as a complex category, having at its basis not only the testimony of the senses, but the evidence of thought and emotions. Let's add to that the awareness of its cultural circumstances." Transformations taking place in the process of experience conceptualization also result in an entirely new way of defining the research method (the place of participant observation is taken here by co-participation), and the distant attitude, which previously was "the condition for maintaining objectivity" is replaced, among other things, with "emotional and empathic engagement" (ibid., p. 21). This way of approaching ethnographic experience reveals the specific role of the subject in the process of knowledge creation; as emphasized by Kaniowska, "in contemporary anthropology, gaining research experience has been turned into an effort to experience the studied reality" (ibid., pp. 20-21).[1] It is worth mentioning that these processes are often accompanied by a belief in the 'truthful' dimension of personal experience (both on the part of a researcher and the researched subject), and the conviction of its authenticity. This in turn leads to an attempt to blur the boundary between the research situation (perceived as a kind of 'unnatural'

---

[1] It is worth adding at the margin of these considerations that experience becomes also an important element of shaping the politics of identity in feminism and within the framework of other trends of the so-called oppositional criticism (Gandhi, 1998). Epistemological perspective of subordinate groups is justified by the "fact of separate experiential spaces" (Hasturp, 1995, p. 153) of individuals subjected to oppression. The position occupied by 'subalterns' in the social structure legitimizes knowledge produced by them as knowledge truly objective and free of ideological bias. One of the symptoms of the 'turn to experience' characteristic for many contemporary trends in humanities is, for example, the concept of "emotive knowledge" coined by Alberto López Pulido. In this project, "experience, emotions, sincerity and empathy are not only methodological tools, but they are also mounted in identity politics, since in fact they play the key role in establishing the location of an author and recovery of his/her 'true' identity" (Domańska, 2008, p. 138; cf. also, Songin, 2011).

one), and daily life as something 'unadulterated.' What happens here is the rejection of an attitude based on the distinction between involvement and detachment, and the personal experience of the researcher is being directly included in the creation of knowledge (cf. e.g. Kafar, 2004; 2010; Michoń, 2010; Pietrowiak, 2011). At the same time, this practice gains not so much *epistemological* as *ethical* justification (Kafar, 2010; Pietrowiak, 2011), which to my mind, at least in part, stems from the belief that the impersonal mode of expression provides incomplete account of the events that have occurred in the field, and thus, it is perceived as insincere, forging the experience, as well as devoid of the necessary sensitivity and the empathic insight (cf. Kaniowska, 2010). Autobiographical texts are in fact intimate descriptions of events and considerations of the authors and their struggle with the matter of field experience, which leads to shifting away from the dispassionate language of realistic ethnography. An important feature of these writings is also that they cease treating the studied subject as an informant, which makes the field account a platform for the contact with another human being (cf. Kafar, 2004; Pietrowiak, 2010).[2] It seems that authors seeking support in ethics are at the same time convinced that empathy, co-participation, co-experiencing, 'moral meltdown of horizons' (cf. Kafar, 2004; 2010), enable reaching the truth, which cannot be achieved by the standard tools of scientific cognition.

Let us now examine more closely the structure of texts that can be considered as exemplification of the genre of autobiographical ethnography, which is characterized by the incorporation of descriptions reflecting the experiences and beliefs of the researchers into the anthropological narration (cf. Reed-Danahay, 2001). The subject of my analysis will be two articles: Marcin Kafar's *Wobec wykluczonych: Antropolog w Domu Pomocy Społecznej dla Przewlekle Chorych* (*Towards the Excluded: Anthropologist in the Aid Centre for the Chronically Ill*) (2010) and Kamil Pietrowiak's *Gdzieś pomiędzy: Przestrzeń spotkania (w terenie)* (*Somewhere in Between: Meeting Space) (in the Field)* (2011).

Already the introduction of the article written by Marcin Kafar suggests that we are not dealing with the classic ethnographic text. But are we sure about that? The Author speaks in the first person, thus clearly indicating his presence. The introduction is a description of the first visit to the Nursing Home where the Reader follows the Author-novice, a little insecure and lost. Our attention is captivated, almost from the beginning,

---

2 Significant in this case is the title of the doctoral dissertation of Marcin Kafar, namely, *I the Anthropologist—I, the Human Being: On a Certain Variant of Engaged Anthropology* (the dissertation was prepared at the Institute of Ethnology and Cultural Anthropology, University of Lódź, 2009, under the direction of professor A.P. Wejland).

by the realistic description. The Author makes a detailed report of what he observed (the "iron fence," "eight steps," "big mirror divided into parts," "cardboard chart," "gate that is about a meter high"), what he experienced ("I clench my hand on a metal rod, give it a pull once or twice, until the lock gives in"), and what he thought ("Who are These, in whose hands the Residents entrust the 'days of their lives'? Who are the Residents themselves?"). As Marcin Kafar (2010, p. 203) points out, the reflections contained in the text are a kind of "a counter-discourse imitating thoughts," which is aimed at introducing "additional story dramatization." This sentence clearly suggests the Reader that he/she deals with the Author conscious of his creation of anthropological narratives. The literary devices used here resemble the efforts of Bronisław Malinowski, aimed at making being There more real not only for the Author, but for the Reader as well (cf. Mokrzan, 2010). Just as Malinowski encourages his Readers to "fully surrender to reading, to come aboard together with the Author and to be with him on the Trobriand Islands" (ibid., p. 28), Kafar forces us to walk with him along the corridors of the Aid Center and feel 'with our own skin' the atmosphere of the place. His (and the Reader's) aim is to hear the history of life of the residents, the stories that are significant not only cognitively, but also therapeutically. Opening oneself to the story of the Other is to make our lives better and more valuable, to break the boundaries separating people and to facilitate "searching for answers to the question 'Who are we?'" (Kafar, 2010, p. 213). As the Author writes, "all theories are useless if we forget that in front of us first of all there is a person who is the subject" (ibid., p. 208). This kind of ethnographic research is filled with the call to "fulfill a moral duty 'to be for the other'" (ibid., p. 209); and it is not about the cultural meanings or social constructs, but a more fundamental issue—the "essence of humanity" (ibid., p. 213).

Kamil Pietrowiak chooses a slightly different way, namely, he follows the path of the dialogue, for which the inspiration is the philosophy of Martin Buber, Emmanuel Lévinas and Józef Tischner. It is not only a conversation between two parties, but a dialogue of the (anthropological?) 'I' with the informants, readers, authors of other texts (thus 'You' is plural in this case). The way of presentation, adopted by the Author—as he admits himself—stems from the perceived fatigue with "some, intellectually over-saturated anthropological texts whose primary purpose seems to be hindering or even preventing the Readers from reading them and whose hermetic language rather tends to indicate the weakness than epistemological opportunities of anthropological research" (Pietrowiak, 2011, p. 26); Pietrowiak, on the other hand, wants to speak "as a normal human being," in order to give pleasure to the Reader (ibid., pp. 26-27). Although the text refers to the fieldwork methodology, the Author conceives it to be something more—it is to be a reflection over a meeting and a conversation of two people. It might also be concluded that this is a description of field experiences that becomes a contribution to the quest for the truth. It is not, however, about the truth (cognition) of the reality, but the veracity of experiencing this reality by the researcher and the researched. But for the promise of the truth that is rendered by experience, "it would be better to sit at home, read more books, watch rather than listen, write texts about texts" (ibid., p. 28).

Ethnographic experience, described by Kafar and Pietrowiak, is examined by both these Authors from the point of view of ethics, and—as Lévinas argues—it involves direct concentration on the Other. The responsibility for the Other is, as the philosopher thinks, "non-removable in its ethical possession. It is the responsibility from which one cannot escape, and thus it becomes a principle of absolute individualization" (Lorenc, 1998, p. 48). In the cited works the relation I—the Other is understood metaphorically as "the cradle of the Real Life" (Buber, 1958, p. 9). Adopting this perspective results in the rejection of the distance and giving the priority to the bond forming between two people; to use the words of Martin Buber, "the one primary word is the combination I—Thou" (Buber, 1958, p. 3). As a result, the anthropological narrative is transformed into a description of our own experience and the one of the encountered people, it is the story of life, stemming from a desire to establish a genuine relationship based on reciprocity and co-experiencing.

The discussed texts induce us to ask the following question: how do the circumstances (experiencing the loss—Marcin Kafar) and the object of study (faith and holiness—Kamil Pietrowiak), determine the narrative and research perspective of anthropologists? Reading these publications also raises doubts as to whether the necessary condition for the Reader to participate in the (fieldwork?) experience of the Authors is sharing their view on the world and a certain type of sensitivity represented by them. In other words, should the reading of these texts not be preceded by an annotation included by Rudolf Otto in his book *The Idea of the Holy*: "Whoever cannot do this, whoever knows no such moments in his experience, is requested to read no farther" (Otto, 1958, p. 8). Certainly, it would be wrong to assume that longing for contact with another human being, so close to Kafar and Pietrowiak, is just as close to all researchers.

In the second part of my paper I will discuss the way of conceptualizing ethnographic experience by female researchers associated with the poststructuralist and feminist approach. However, I would like to stress that the perspective adopted by Kafar and Pietrowiak, as well as those presented below, I consider as different, but in equal extent legitimate forms of reflection on the specifics of the field research and the role of the subject in the generation of anthropological knowledge.

CB

One of the risks associated with the adoption of the perspective favored by Kafar and Pietrowiak is, in my opinion, the assumption that the cognitive process is subjective in nature and it is not focused on the subject of study, but—to use the words of Roy D'Andrade—on how an anthropologist doing the description of the reality responds to or reacts to the object of the description (D'Andrade, 1995, p. 399). Such a conviction stems from a misinterpretation of the consequences arising from the criticism of objectivism. In the positivist model of research, knowledge-building is defined as a process from which there are eliminated any and all personal factors. Hence, the logic of this process is based on dichotomies such as: 'objective—subjective,' 'rational—emotional,' 'mental—physical,' 'personal—professional,' 'intuitive—analytical.' The experience of the subject is located here in the personal, subjective or private sphere, and lies outside of the actual area of research ('professional') interest of an anthropologist. The criticism of objectivity and rationality, however, should not be identified with the validation of subjectivity and emotions as the dominant tools of acquiring knowledge, because they, standing in opposition to the Enlightenment strategies constituting the authority of the subject, are included in the Enlightenment framework of the discourse (cf. Bar On, 1992). To contribute to a better understanding of the reality studied by anthropologists, reflexivity must, to my mind, become an effort oriented on going beyond the categories embraced in the positivist model of knowledge construction. In the execution of this task, one of the helpful concepts is that of Donna Haraway's "situated knowledges," which refers to "politics and epistemology of location."[3] The knowledge of the subject, resulting from its situatedness, is considered here in terms of a kind of semiotic-material technology linking the bodies and meanings. This means that while trying to understand ourselves, we do it in a symbolic language mode, we narrativize our own experience. Haraway's proposal allows for casting away the understanding of the subject as identical with itself, reductionist and transparent at the same time (cf. Haraway, 1988).[4] Experience is perceived

---

[3]   A similar value has the concept of the so-called strong objectivity coined by Sandra Harding. The author points to the need to disclose the history, location, influences, beliefs and moral views of a researcher at every stage of the research project. In other words, the researcher is required to continuously disclose his/her standpoint throughout the duration of the project (cf. Harding, 1991; 1993).

[4]   It is interesting to note that the problem of the situatedness of the subject is also present in the positivistic research model. The necessity to eliminate the evaluative judgments assumes *implicite* its location.

in a similar way by Joan Scott, who emphasizes that it must be regarded as an event of a discursive nature. According to her, there are no "individuals who have experience, but subjects who are constituted through experience." Thus, in her opinion, what can be seen and what can be felt is not an "evidence that grounds what is known," but should rather be problematized as something that requires further review and analysis (Scott, 1992, p. 26). Scott claims that "experience is always a constructed category that contains the ideological traces of the context from which it emerges" (Domańska, 2008, p. 135; cf. Scott, 1991; see also, Scott, 1992). Thus, experience is not a tool, which makes it possible to reach directly some kind of external, non-ideological reality. The adoption of an opposite assumption would lead to the conclusion that the identities of the researcher and the researched are self-explanatory. Scott strongly emphasizes that situatedness must be understood as a place of intersection of various discourses taking an active part in the forming of subject positions (Scott, 1992, p. 25).[5] The reasoning used by both researchers argues that reflective consideration over the role played by experiences in the process of knowledge production cannot be limited to describing the private feelings and views of the researcher and his/her informants. Due to such practice, anthropologists risk being accused of solipsism and narcissism. In addition, it is also questionable in terms of cognition.

Valuable tips, regarding the potential uses of reflection on the meaning of subjective experience in the production of knowledge, can be found in the texts of feminist scholars.[6] They point to the role played in research by different kinds of discursively constructed identity categories. The factors emphasized by them include historical, national and generational factors, as well as race, class, gender and sexuality, taking an active part in the formation of subject positions. According to feminists, it is acceptable and even recommended to consider "the aspect of being as a way of knowing" (Wickramasinge, 2006). Situatedness allows one to understand the dynamic nature of the mutual interactions between the different identity categories that determine the ways of interpreting the reality. In the opinion of Donna Haraway, situatedness is not static, defined or fixed, on the contrary—it is relational and unstable, which results from its contextual

---

[5]   The poststructuralist approach assumes that a subject is a function of discourse.

[6]   The debates conducted by radical anthropologists in the 1970s were also of great importance for the development of the interest in this subject. This refers primarily to the reflections on the so-called native knowledge (Jones, 1970) and perspectivistic knowledge (Lewis, 1970), the influence of ideological factors, as well as the impact of the 'markers of identity' such as culture, class or nationality on the way of constructing knowledge in anthropology.

nature (Haraway, 1988). This approach emphasizes the diversity of subjective locations and points of view, encouraging feminist researchers to reflect on their situatedness in research projects. It can therefore be concluded that location is the exemplification of the active inclusion of the 'self' in the process of knowledge production, which in feminism is considered from the point of view of politics, understood as a synonym of strategy and means "critical reflection on the consequences of our own location in the world, the values in which we believe, the objectives we are trying to achieve [...]; power relations, in which everyone, in one way or another, is involved" (Baer, 2005, p. 7). In short, this approach suggests that experience is created as a result of the subject's situatedness in the social and cultural discourses and thus it cannot be comprehended in terms of subjectivity.

Feminist anthropologists problematize ethnographic experience and sensitize us, *inter alia*, to the perceptions of a researcher and her femininity by the researched. Peggy Golde (1986) indicates that women in the course of fieldwork tend to be treated as androgyny, honorable men, children or as weak creatures that need constant care and protection. Sometimes it also happens that in order to be able to participate in the local community life, they must pass a series of initiation rituals by which they acquire a symbolic identity, defining their place within the culture they study. Carol Warren (1988), in turn, argues that female anthropologists usually have a lower status in the researched communities and their authority is created on the basis of their race, class or culture. The studies conducted by Elizabeth Enslin show, however, that the distinction such as: 'the self' and 'the Other' or distance and commitment, are now unnecessary and often unfounded. Enslin, who is an American and the wife of an Indian anthropologist (a graduate of Oxford University), carried out a research in her husband's home village. Her interests were mainly focused on the problems of landless women. In addition to family connections, another difficulty in the research project was the fact that the relatives of Enslin's husband—taking up an activist action for the creation of the Help Center for Women—required that Enslin should engage in local politics. The situation resulted in a serious dilemma concerning the boundary between her and the Others, who due to her marriage became a part of her family. This dilemma, in a broader horizon, concerned the need for differing the levels of engagement and distance (Enslin, 1994).

Reflection on the situatedness of the female subject facilitates distinguishing a lot of interdependencies important from the point of view of the research process and the cultural ways of defining gender. One of the strategies used during the field research is following the gender dos and don'ts specific for a given cultural area. Female researchers use in this way the perspective of 'a look from within' underpinned by the concept of "embodied subjectivity" (Smith, 1987).[7] On multiple occasions, the mere fact of being a woman compels a researcher to submit to the local norms in a much more severe way than in case of male anthropologists (Wolf, 1996). Diane Wolf reminds us of the difficulties faced especially by those anthropologists who led their studies in societies with strong partiarchalism, to which they often gained access thanks to the privileged position of certain men: their fathers, husbands or brothers (Oboler, 1986; Berik, 1996). Feminists carrying out research projects in the Middle East or South Asia point to the necessity of wearing traditional local clothes and complying with certain rules such as the prohibition of looking at men or talking to them in certain situations (Wolf, 1996; cf. Pettigrew, 1981; Abu-Lughod, 1986; Schenk-Sandbergen, 1992). Similar restrictions apply to women conducting research in caste communities, where female anthropologists are required to act in accordance with the established cultural rules. They cannot perform activities traditionally associated with different spheres of the society, such as cleaning their own homes or toilets, cooking and eating food with people coming from certain castes (Wolf, 1996; cf. Kumar, 1992; Schenk-Sandbergen, 1992).[8] Unmarried female anthropologists face various pressures on the part of the researched with regard to changing their social status or must be prepared for the fact that they may become addressees of marriage proposals. Married women, in turn, may be advised that the proper place for them is not 'in the field,' but at home with their children (Enslin, 1990; after Wolf, 1996, p. 9). Female researchers also found themselves in situations where if they were pregnant, the status of impure women was attributed to them (Enslin, 1990).

Feminist anthropologists clearly show that knowledge is produced by subjects having gender, nationality, sexuality and age, and although they acknowledge that "the autobiography of fieldwork is about lived interactions, participatory experience and embodied knowledge" (Okely, 1992, p. 3), this does not mean that it does not require in-depth theoretical consideration; as stated by Ewa Domańska (2008, p. 131), "experience (along with other concepts accompanying it such as memory, testimony, emotions, trauma, empathy and compassion) belongs to 'engaged' categories, which require particular vigilance."

---

[7] This approach enables including the researcher's own experiences and beliefs in the process of knowledge production.

[8] It is worth indicating that these restrictions apply equally to male and female anthropologists.

In my opinion, ethnographic experience, although inevitably linked to what is individual, should stay far away from the practice which aims at describing personal experiences of the researcher and the researched, because, as instructed by Cifford Geertz (2000, p. 58), "The trick is not to get yourself into some inner correspondence of spirit with your informants." Recognition of the research process as a "complex intercultural mediation, and a dynamic interpersonal experience" (Scholte, 1974, p. 438), allows to conclude that experience is indeed born somewhere "between us and the Others" (Hastrup, 1987). Special importance of the word 'between' must be highlighted here, as it indicates both an important mediating role of a language and the fact that 'the self' is always relational and it is defined in relation to the Other. Therefore, experience does not correspond to the category of authenticity, since to become understandable to a researcher (or the researched) it must be viewed from a distance. In this way, the desire to directly approach the Other, associated with attempts to erase the border between *life* and *research*, can never be fulfilled. The effort oriented on diminishing the distance between the self and the Other is dissipated in the process of understanding, which inevitably assumes at least the minimum narrative distance (cf. Songin, 2010, pp. 77-80).

References

Abu-Lughod, L. (1986). *Veiled Sentiments: Honor and Poetry in a Bedouin Society.* Berkley: University of California Press.

Baer, M. (2005). Ku pluralistycznej wspólnotowości. *(op.cit.,). Maszyna interpretacyjna. Pismo kulturalno-społeczne,* 6 (27), 6-7.

Bar On, B. (1993). Marginality and Epistemic Privilege. In L. Alcoff & E. Potter (Eds.), *Feminist Epistemologies* (pp. 83-100). Series: Thinking Gender. New York – London: Routledge.

Berik, G. (1996). Understanding the Gender System in Rural Turkey: Fieldwork Dilemmas of Conformity and Intervention. In D.L. Wolf (Ed.), *Situating Feminist Dilemmas in Fieldwork* (pp. 56-71). Boulder: Westview Press.

Brocki, M. (2008). *Antropologia: Literatura—Dialog—Przekład.* Wrocław: Wydawnictwo Katedry Etnologii i Antropologii Kulturowej Uniwersytetu Wrocławskiego.

Buber, M. (1958). *I and Thou.* (R.G. Smith, Trans.). New York: Charles Scribner & Sons.

Buchowski, M., & Kempny, M. (1999). Czy istnieje antropologia postmodernistyczna? In M. Buchowski (Ed.), *Amerykańska antropologia postmodernistyczna* (pp. 9-28). Warszawa: Instytut Kultury.

Burszta, W. (1992). *Wymiary antropologicznego poznania kultury*. Poznań: Wydawnictwo Naukowe Uniwersytetu Adama Mickiewicza w Poznaniu.

D'Andrade, R. (1995). Moral Models in Anthropology. *Current Anthropology, 36* (3), 399-408.

Domańska, E. (2008). Doświadczenie jako kategoria badawcza i polityczna we współczesnej anglo-amerykańskiej refleksji o przeszłości. In A. Zeidler-Janiszewska & R. Nycz (Eds.), *Nowoczesność jako doświadczenie: Dyscypliny—paradygmaty—dyskursy* (pp. 130-142). Warszawa: Wydawnictwo Szkoły Wyższej Psychologii Społecznej Academica.

Enslin, E. (1994). Beyond Writing: Feminist Practice and the Limitations of Ethnography. *Critical Anthropology, 9* (4), 537-568.

Enslin, E. (1990). *The Dynamics of Gender, Class, and Caste in Women's Movement in Rural Nepal*. Doctoral dissertation, Department of Anthropology, Stanford University.

Gandhi, L. (1998). *Postcolonial Theory: A Critical Introduction*. New York: Columbia University Press.

Geertz, C. (2000). *Local Knowledge: Further Essays in Interpretive Anthropology*. New York: Basic Books.

Golde, P. (1986). Introduction. In P. Golde (Ed.), *Women in the Field: Anthropological Experiences* (pp. 1-19). Berkeley: University of California Press.

Haraway, D. (1988). Situated Knowledges: The Science Question in Feminism and the Privilege of Partial Perspective. *Feminist Studies, 14* (3), 575-599.

Harding, S. (1993). Rethinking Standpoint Methodology: What is Strong 'Objectivity'? In L. Alcoff & E. Potter (Eds.), *Feminist Epistemologies* (pp. 49-82). Series: Thinking Gender. New York – London: Routledge.

Harding, S. (1991). *Whose Science? Whose Knowledge? Thinking from Women's Lives*. Ithaca: Open University Press.

Hasturp, K. (1995). *A Passage to Anthropology: Between Experience and Theory*. London – New York: Routledge.

Hastrup, K. (1987). Fieldwork among Friends: Ethnographic Exchange within the Northen Civilisation. In A. Jackson (Ed.), *Anthropology at Home* (pp. 94-108). New York: Tavistock/Methuen.

Jones, J.D. (1970). Toward a Native Anthropology. *Human Organization, 29*, 251-259.

Kafar, M. (2010). Wobec wykluczonych: Antropolog w Domu Pomocy Społecznej dla Przewlekle Chorych. In K. Kaniowska & N. Modnicka (Eds.), *Etyczne problemy badań etnograficznych* (pp. 203-227). Wrocław – Łódź: Polskie Towarzystwo Ludoznawcze.

Kafar, M. (2004). Od spotkania do wspólnoty: Autobiograficzny raport z trenu. In G.E. Karpińska (Ed.), *Codzienne i niecodzienne: O wspólnotowości w realiach dzisiejszej Łodzi* (pp. 79-101). Łódź: Polskie Towarzystwo Ludoznawcze.

Kaniowska, K. (2010). Skąd się biorą etyczne problemy badań antropologicznych? In K. Kaniowska & N. Modnicka (Eds.), *Etyczne problemy badań etnograficznych* (pp. 7-16). Wrocław – Łódź: Polskie Towarzystwo Ludoznawcze.

Kaniowska, K. (2006). Dialog i interpretacja we współczesnej antropologii. *Etnografia Polska, 50* (1-2), 17-34.

Kaniowska, K. (1999). *Opis—klucz do rozumienia kultury*. Wrocław – Łódź: Polskie Towarzystwo Ludoznawcze.

Kumar, N. (1992). *Friends, Brothers and Informants: Fieldwork Memoirs of Banaras*. Berkley: University of California Press.

Lewis, D. (1973). Anthropology and Colonialism. *Current Anthropology, 14* (5), 581-602.

Lorenc, W. (1998). *W poszukiwaniu filozofii humanistycznej: Heidegger, Lévinas, Foucault, Rorty, Gadamer*. Warszawa: Wydawnictwo Naukowe Scholar.

Malinowski, B. (1967). *A Diary in the Strict Sense of the Term*. (V. Malinowska, Pref.; R. Firth, Intr.; N. Guterman, Trans.). London: Stanford University Press.

Michoń, Ł. (2010). Spowiedź. *Tematy z Szewskiej, 1* (4), 83-88.

Mokrzan, M. (2010). *Tropy, figury, perswazje: Retoryka a poznanie w antropologii*. Wrocław: Wydział Nauk Historycznych i Pedagogicznych Uniwersytetu Wrocławskiego, Katedra Etnologii i Antropologii Kulturowej.

Oboler, S.R. (1986). For Better and Worse: Anthropologists and Husbands in the Field. In T.L. Whitehead & M.E. Conaway (Eds.), *Self, Sex and Gender in Cross-Cultural Fieldwork* (pp. 28-51). Urbana: University of Illinois Press.

Okely, J. (1992). Anthropology and Autobiography: Participatory Experience and Embodied Knowledge. In J. Okely & H. Callaway (Eds.), *Anthropology and Autobiography* (pp. 1-28). London – New York: Routledge.

Otto, R. (1958). *The Idea of the Holy: An Inquiry into the Non-Rational Factor in the Idea of the Divine and its Relation to the Rational*. (J.W. Harvey, Trans.). Oxford: Oxford University Press.

Pettigrew, J. (1981). Reminiscences of Fieldwork among the Sikhs. In H. Roberts (Ed.), *Doing Feminist Research* (pp. 62-82). Boston: Routledge & Kegan Paul.

Pietrowiak, K. (2011). Gdzieś pomiędzy: Przestrzeń spotkania (w terenie). *Tematy z Szewskiej, 1* (5), 25-34.

Reed-Danahay, D. (2001). Autobiography, Intimacy and Ethnography. In P. Atkinson, M. Coffey, S. Delamont, J. Lofland, & L. Lofland (Eds.), *Handbook of Ethnography* (pp. 407-422). London – Thousand Oaks – New Delhi – Singapore: Sage.

Schenk-Sandbergen, L.C. (1992). *Gender in Fields Research: Experiences in India.* Occasional Paper, IDPAD, the Netherlands.

Scholte, B. (1974). Toward a Reflexive and Critical Anthropology. In D. Hymes (Ed.), *Reinventing Anthropology* (pp. 430-457). New York: Pantheon Books.

Scott, J. (1992). Experience. In J. Butler & J.W. Scott (Eds.), *Feminist Theorize the Political* (pp. 22-40). New York – London: Routledge.

Scott, J. (1991). The Evidence of Experience. *Critical Inquiry, 17* (4), 773-797.

Smith, D. (1987). *The Everyday World as Problematic.* Toronto: University of Toronto Press.

Songin, M. (2011). Z podporządkowanego punktu widzenia: Roszczenia poznawcze klas podrzędnych. In A. Malewska-Szałygin & T. Rakowski (Eds.), *Humanistyka i dominacja: Oddolne doświadczenia społeczne w perspektywie zewnętrznych rozpoznań* (pp. 29-46). Warszawa: Instytut Etnologii i Antropologii Kulturowej Uniwersytetu Warszawskiego.

Songin, M. (2010). Etyczny wymiar spotkania etnograficznego w świetle antropologii zaangażowanej. In K. Kaniowska & N. Modnicka (Ed.), *Etyczne problemy badań etnograficznych* (pp. 69-86). Wrocław – Łódź: Polskie Towarzystwo Ludoznawcze.

Van Mannen, J. *(1988). Tales of the Field: On Writing Ethnography.* Chicago: University of Chicago Press.

Warren, C. (1988). *Gender Issues in Field Research.* Newbury Park: Sage.

Wickramasinge, M., (2006). An Epistemology of Gender: An Aspect of Being as a Way of Knowing. *Women Studies International Forum, 29,* 606-611.

Wolf, L.D. (1996). Situating Feminist Dilemmas in Fieldwork. In D.L.Wolf (Ed.), *Feminist Dilemmas in Fieldwork* (pp. 1-55). Colorado: Westview Press.

Chapter Eight

FROM THE POINT OF VIEW OF A WOMAN

AUTOBIOGRAPHICAL CONTEXT OF FEMINIST STUDIES

by Aneta Ostaszewska

> The real courage begins when we refuse
> to treat life as a series of missed opportunities...
>
> E. Cioran, *Solitude et destin*

Writing this text I am attempting to reflect on the autobiographical dimension of my research work. I am trying to express this reflection *in extenso*, as honestly as I can. I will start with writing in the first person. In this way, I am trying to achieve symbolic empowerment of myself with a view to carry out self-identification as a social researcher, and also a feminist.

My academic achievements are thematically very diverse. I have written on various topics: pop culture, narcissism, taboo... And all this was born out of me (to paraphrase Adrienne Rich),[1] but I have to admit that I do not feel the satisfaction of the published books and articles. On the contrary, I have always had a sense of 'insufficiency' concerning words, some unsaturation, fragmentation. I have longed for something and it has been the kind of longing which makes one sad, because it results from knowing that you can (you could) avoid this yearning, but for the lack of courage. Today—to a significant degree thanks to my readings[2]—I have

---

[1] I refer to the title of a book by Adrienne Rich, *Of Woman Born* (1986).

[2] My most important recent readings have been the essays by, among other authors, bell hooks, Carolyn Ellis and the aforementioned Adrienne Rich. An important source of

enough courage to admit that I was not convincing enough for myself in what I did; and I was not convincing because I did not realize the potential that I have; that is, I did not use the potential of my autobiographical experience. It comprises, first of all, the experience of being a woman; a woman at the university, a sociologist, a researcher. These experiences are important because they determine my sense of identity.

Write Myself, Write with Myself

Despite the fact that for several years I have been dealing with sociological and anthropological analysis of the phenomena of contemporary culture, I have not written *explicite* about experiencing the reality (social, academic) from the point of view of a woman; a woman who wants to express something with her own voice, but instead of speaking she remains silent or talks about something else. I denied myself the right to speak of what I wanted to say. I was looking for 'substitutional' subjects, I imitated the language foreign to me, I learned from others, carefully studying their words, thoughts, and biographies—all in order to get confirmation that what I'm doing meets the criteria of 'scientificity';[3] not wondering at all whether or not it is possible to have science existing independent of the knowing subject (Smith, 1990, p. 62).[4] I shaped my 'professional' biography as well as my research and writer workshop not based on my own experience, but on the knowledge acquired from others, on the methods coined by others and on their mistakes. And one should write 'his/her way,' i.e. write without forgetting about who he/she is. Who am I?

I do not hesitate to place this question (questions are a condition of self-reflexivity), although I am aware of the pathetic tone, which can be caused by the further reflection moving forward in the direction of the most fundamental issues—existential and decisively undecidable. However, the question, 'Who am I?' makes me wonder not so much about the specific statements and visions of my own identity, but the means of

inspiration have been (and are) the biographies of women, including Simone Weil and Hannah Arendt.

[3] I used quotation marks because I had (and have) doubts as to how scientificity is defined in social sciences. I wonder whose this scientificity is, since it is not based on the researchers' reflection on the correctness and authenticity of their own work, but from the beginning—as part of the socialization of the university—it has been the result of uncritically inculcated abstract rules, imposed from the outside, without regard to the contexts: situational, historical, political, etc.

[4] I refer here to a question posed by Dorothy Smith, "How can there be 'knowledge' that exists independently of knowers?" (Smith, 1990, p. 62).

looking for/discovering tips/answers, ways of documenting the process of 'becoming myself.' One of these ways is writing. Thus, in response to the question, 'Who am I?', I can at this point say: 'The one who writes.' What does she write? About what does she write?

Through ethnographic and autoethnographic inspirations I know that one can write *with* experience and *about* experience (Ellis, 2008; 2004; Ellis & Bochner, 1996). And this is my challenge: to make the autobiographical episodes the source of reflection and inspiration in my work on biographical experiences of women. I would like, using (auto)biographical experience—my own and that of other women—to perform a reflection on the dynamics and contextuality of identity. I am also interested in the language through which the experiences getting into the realm of words become a 'text,' and at the same time are granted the intersubjective availability. Simultaneously, at the time of being articulated and subjected to reflection, these experiences are (re)constructed and lived.

Laurel Richardson (2000, p. 925) notes that "The researcher—rather than the survey, the questionnaire, or the census tape—is the 'instrument.' The more honed the researcher, the better the possibility of excellent research." Therefore, I would like to reach out to others, coming out of myself. To this end, I intend to use writing as a specific method of research, and myself as a tool. My ambition is to write on a road to self-discovery, emancipation (including intellectual and 'professional' emancipation); to write from the point of view of a woman, and more specifically: a woman identifying with feminism.

Feminism—in a general, broad sense—I understand as a social philosophy and political movement;[5] first of all questioning the androcentric social order (disagreement to defining male as universal and neutral), and second: postulating to replace the patriarchy and male domination with gender equality.[6] In a narrower, more personal sense, feminism is for me the question of the modus of life—a life lived in an autonomous and reflective way; this is a question backed up by the courage to talk about oneself in terms of the subject, as well as to pursue the realization of one's own potential (cf. Braidotti, 1995, p. 34). My feminism I would include into the "third-wave feminism" (Starr, 2000, p. 474), that is feminism

---

[5] My understanding of the term 'feminism' refers primarily to the tradition of the 'second-wave feminism,' whose origins date back to the 1960s of the twentieth century.

[6] The main issues are: equal civil rights, equal rights to work and adequate remuneration, equal political rights, the right to dispose of one's own body, equal access to education. What seems to me to be of particular importance is the feminist demand for the access to science, both when it comes to education and 'pure' knowledge (the possibility of creative work, reading, writing, etc.).

growing out of the criticism of maternal inheritance (the 'second-wave feminism') as a starting point to determine its own identity makes negative identification and is therefore a manifestation of individualism opposed to (forced) community (Świerkosz, 2010).[7] The process of individuation of a woman feminist is seen in this model as a result of her mis-identification (or an ambivalent relationship) with the symbolic mother and sisterhood, and with other women. It is about the kind of 'disconnection' or 'non-belonging,' which is not the result of exclusion, but voluntary action, and which creates a kind of nomadic subjectivity (Braidotti, 1994). The 'third-wave feminism'—which I think is particularly important—uses the description of the individual experiences of a woman and applies for this purpose the genre of autobiographical essay (Graff, 2003).

Feminist Methodology

Making the reflection on the autobiographical experience an inherent part of my research work, I support—which has been signaled above—a feminist perspective. I'm interested in the reflection carried out using the feminist categories, both in relation to the issues of epistemology and methodology. Behind this decision there are some valid arguments, whose presentation I choose as the main objective of this chepter.[8] These arguments stem from the following questions: 'What can be offered by the methodology oriented in this way?' and 'What does the feminist perspective give me—a particular researcher?'

An attempt to answer the above questions should start with explaining what feminist methodology is. For this purpose, first of all, I reach for the texts that have already become classics (from the late 1980s and 1990s of the twentieth century), including those by Sandra Harding,

---

[7] 'Third-wave feminism' is an umbrella term for a variety of events, attitudes, and practices related to the feminist movement of the 1980s of the twentieth century. We can distinguish three main aspects in it: 1. Emphasis on the problems of women in the least industrialized countries in the world; 2. Criticism of the values dominant at work and in the society; 3. Eliminating barriers to experiencing love and sexual pleasure for women.

[8] The pedagogical purpose—so to speak—would be to nullify the kind of aversion to anything that is associated with the term 'feminism.' Every time I try to talk about feminist methodology, I come across rather significant reactions. These are usually not so much critical voices as such trying to discredit the methodology practiced by scholar-feminists. What is hidden in these statements is a message not expressed directly, but implying the futility of the research perspective appointed by feminism. It regards more the reservations and doubts about feminism that causes specific negative connotations, than the methodology itself.

Donna Haraway and Dorothy Smith. The texts of these authors not only contributed to the overall interest in the feminist methodology, but also designated the direction of the criticism of that methodology, both outside and within the feminist movement.[9]

The major problem associated with the settlement of what feminist methodology is lies in the fact that even among the feminist-oriented researchers themselves there is no consensus as to one common definition (Ramazanoğlu & Holland, 2002, p. 8).[10] Sandra Harding (1987a) asks bluntly, 'Is it a particular method, a set of methods, or a research strategy?' This question remains open, similarly to another one: 'Is it even possible (and advisable) to talk about a methodology common to all feminist researchers?' Jennifer Brayton argues that the ambiguity of structure of the feminist research is the reflection of the ambiguity of feminism as a theory and practice (what seems to be common to the different strands of feminism is the general focus on the category of 'gender' (Brayton, 1997)). Virginia Olesen (2005, p. 236) explains the diversity of the forms of feminism in the following way: "Feminisms draw from different theoretical and pragmatic orientations that reflect national context where feminist agendas differ widely." As it turns out, feminism takes on many forms and it is a dynamic notion subjected to the (continuous) process of *rewriting*.[11] In this situation, it seems to be appropriate to display only a certain set of the generally accepted rules that govern the research conduct.[12]

---

[9] The preparation of this text—within the elaboration of the primary sources—ran in a chronological order (from the oldest texts to the newest ones). More attention has been devoted by me to classical texts (which contributed to the emergence of feminist methodology) than to the works being a later development, revision and criticism of the primary sources. The discussion of the latest publications on feminist methodology will be the subject of a separate article.

[10] It would be more correct to use the term 'feminist methodologies,' but the literature of the subject has adopted the form of a single 'feminist methodology' (this does not preclude the understanding of the term as a set of different research orientations). This measure itself seems above all to emphasize the close relationship with feminism, and the secondary issue is the complexity and scope of the definiendum. Caroline Ramazanoğlu and Janet Holland argue that the use of the term 'feminist methodology' is a short-cut and it mainly serves maintaining the simplicity of terminology. This way of comprehension I also adopt for the purpose of this discussion.

[11] I realized this while writing this article. The multiplicity of approaches and interpretations of what feminism is (from the perspective of the twenty-first century) calls for a synthetic collective recognition, but at this stage of my research, this task turns out to be extremely difficult and requires more substantial preparation.

[12] My aim is not to present a comprehensive review of feminist research strategies, but to identify several key features, the most important ones from the point of view of the tackled problem, namely the autobiographical contexts of research.

Mary Maynard (1995, p. 106) considers as an important designatum of the feminist methodology "the question we have asked, the way we locate ourselves within our questions, and the purpose of our work." Brayton (1997), in turn, indicates that what constitutes the uniqueness of feminist research are the motives, interests, and knowledge brought into the research process.

Searching for the subsoil supporting the emergence of feminist methodology, without much risk we can accept that what worked here was a mechanism to resist the dominant models of research, previously functioning within social sciences. Their use not only proved to be rather ineffective (no answers to the posed questions), but also revealed the illusory nature of the preconceived assumptions, including those on the neutrality and objectivity of the researcher (Ramazanoğlu & Holland, 2002, p. 8). The traditional (sociological) strategies, although they made females the subjects of research, did not take into account their point of view, neglected their experiences, needs, social roles, i.e. all that would make women the actors in the research. Simply put, studies run according to the conventional model produced a male vision of femininity; confirmed (men's) vision of the world in which there was a breakdown by gender, in which the roles were rigidly defined, and the position of women was confined to the place designated by the dominant social order.

Most commonly, the term 'feminist methodology' is associated with the research carried out *by* women, *for* women and *with the participation of* women. It is true that feminist research is focused around the female theme and problematizes different situations of women (the studies include, among other things, the material conditions of women's lives, their status and social roles; relationships with men, children and other women; attitudes towards their own bodies, the problem of identity and subjectivity; and finally: the relationship with their own history). The exploration covers the networks of social relationships between gender and economic forces, family, sexuality, politics, science, etc. The purpose of this research is to create equal opportunities for women to express themselves and to participate in the academic, social and political life. The accent is put on the exploration (listening to) of the things women have to say. Mary Maynard (1995, p. 103) explains, "What was most usefully required was an approach to research which maximized the ability to explore experience, rather than impose externally defined structures onto women's lives. Thus feminists emphasised the importance of listening to, recording and understanding women's own description and accounts." For this reason, feminist studies resign from the introduction of pre-categorization. Maynard justifies this decision as follows: "At its heart was the tenet that feminist research must begin with an open-ended exploration of women's experiences, since only from that vantage point is it possible to see how their world is organized and the extent to which it differs from that of men" (ibid., p. 103). The research pursuit is therefore—generally speaking—mining and understanding the meanings which women attribute to their experiences.

Questions that emerged very quickly from the feminist research were: 'What is meant by a 'woman'?' 'When I say a «woman», do I mean the biological sex or a set of some characteristics attributed to women?' 'In whose name do I actually speak and can I, as a feminist researcher, speak «on behalf of» (all) women?'[13]

The claim that (all) women form one community (due to their gender) share the same social positions that can be studied by a researcher (also a woman) who 'understands' them is incorrect.[14] Women do not have the same experiences and do not form single subjectivity. Caroline Ramazanoğlu and Janet Holland (2002, p. 8) stipulate, "Feminists cannot speak for 'we humans,' 'we women' or 'we feminists' without specifying the nature and boundaries of the collectivity or category they speak for." Therefore, as it turns out, the category 'woman' is what conditions the complexity of the feminist research. The fact that these studies are done by women and including women cannot be a distinctive trait of feminist research. Gender alone does not guarantee access to knowledge and life of a woman (Olesen, 2005, p. 248). It is rather about not ignoring the differences that exist between women, which relate to both the individual experiences of women and different socio-cultural features (social class, ethnicity, etc.). Olesen explains, "As the concept of a universalized woman or women faded, understanding grew that multiple identities and subjectivities are constructed in particular historical and social contexts" (ibid, p. 241). The focus, therefore, is pinned on the exploration and analysis regarding the reality of the position of individuals or groups, with a strong emphasis on the diversity and differences of biographical experiences. Ilene Alexander points out that feminist thought has many ways

---

[13] These questions are very important to me because they cause a series of autobiographical references. My process of individuation as a woman ran in the context of a (strong) opposition to other women and in the general state of loneliness. Particularly painful was experiencing Otherness towards/among women. My feminism was born in opposition not only to the structures of patriarchy, but also to other females, including feminist units organized as a group of women whom I did not want to identify with.

[14] I am skipping at the moment (as falling outside the scope of this text) the discussion on the crisis of representation and legitimacy in the contemporary humanities and social sciences.

of defining the differences; she clearly emphasizes that these are "differences" and not "divisions" (Alexander,1989, p. 100). It can be therefore assumed that the subject of interest of feminist methodology are different experiences of different women. But only women?

An important finding of the 'second-wave feminism' were the categories of a 'difference' and 'gender.' The concept of the difference, in a broad, general sense, refers to the (dichotomous) division of the world into masculine and feminine. However—according to Sandra Lipsitz Bem (1993, pp. 2-3)—the difference is more than a guiding principle regulating and organizing the social life.[15] The difference conditions the existence of a strong relationship between gender and every aspect of life (styles of dressing, speaking, experiencing emotions, etc.). Gender is the sociocultural sex, understood as a social construct, the relation of the difference between the biological sex and the cultural sex (Chodorow, 1995; Brannon, 2002; Harding, 2008, pp. 110-114). This term refers to features, social roles, behaviors, stereotypes, etc., i.e. a complex set of elements related to the construction of gender (the creation of images, expectations and ideas about gender) in a given society. Gender intersects with the class, racial, sexual, ethical, and regional modalities. The scope and multithreading of this concept result in a situation in which feminism comprises many theoretical and research approaches of this category. Olesen (2005, p. 250) says that "Gender, the workhorse concept of feminist theory and research, also has undergone changes that make contemporary use of this concept much more complex and differentiated than at the outset of the second wave." Consequently, Caroline Ramazanoğlu and Janet Holland propose to use in place of 'gender' the term 'gendered lives,' which—since they consider it to be more accurate—refers to the research concerning the differences and similarities of individual human biographies, relationships, inequality and experiences (Ramazanoğlu & Holland, 2002, p. 6). In this way, rather than on gender, we should focus on the specific sub-categories, such as for example: power relations; sexuality and reproduction; sexual differences; the social constructs of male/masculinity and female/femininity; relationships and social interactions; social practices and discourses; corporeality. Clarifying the answer to the question regarding the subject of interest of

---

[15] Sandra Lipsitz Bem talks about the polarization of kinds, which she includes, together with androcentrism and biological essentialism, to the prisms of a kind, that is, the schemes of thinking, responsible for the preservation of the existing, unequal social order. These prisms, according to Bem, reproduce the male domination in two ways: the cultural and social practices put women and men on uneven positions; and secondly— people in the process of socialization absorb the patterns imposed on them concerning the reception of the world and shape their identity accordingly.

feminist methodology, we can say that now it embraces 'gendered lives,' various experiences of different people (men and women); the experience of what it means to be a woman/man.

On the basis of the outlined attempts to define what feminist methodology is and what its object of study is, I'm now going to tackle three basic approaches that exist among researchers-feminists. I will refer to the already classic proposal of Sandra Harding, who was one of the first to make the epistemological recognition within the feminist methodology (Harding, 1987b, pp. 182 ff.; 1991, pp. 105-137); Harding distinguished the following variations of feminist research: (i) feminist empiricism; (ii) standpoint theory, and (iii) temporary epistemologies (postmodern theories).

Feminist empiricism is an approach whose core aim is active inclusion of women—as actors and researchers—into science. The presence and scientific research activity of women have influenced the improvement of the quality and reliability of scientific knowledge, for many years limited only to the male perspective. For this purpose, it is important to identify the prejudices inherent to the methodological assumptions underlying various disciplines of science. According to female researchers, androcentrism—responsible for blocking women's access to education and inhibiting their development—can be overcome through two complementary ways, namely: empowerment of the perspectives of women researchers in science and the use of more adequate research methods and principles of methodological correctness (Haraway, 1988; Harding, 1991, pp. 19-50).

The standpoint theory refers to the Marxist thesis of the epistemological superiority of knowledge from the oppressed class; in this case, women are an oppressed class. As Olesen (2005, p. 243) writes, "standpoint research [...] took up the feminist criticism of the absence of women from or marginalized women in research accounts and foregrounded women's knowledge as emergent from women's situated experiences."[16] It is worth noting that these were the advocates of the standpoint theory that "dissolved the concept of essentialized, universalized woman, which was to be replaced by the ideas of a situated woman with experiences and knowledge specific to her place in the material division of labor and the racial stratification systems." Olesen includes sociologists Dorothy Smith and Patricia Hill Collins, political scientist Nancy Hartsock and philosopher Sandra Harding among the main representatives of this trend.

Postmodern theories (the so-called transitional epistemologies) have the form of a loosely associated collection of views; although they refer to

---

[16] In accordance with the standpoint theory, claims to knowledge are socially situated and some social places are better than other as starting points for the study of particular problems and acquiring knowledge.

feminist empiricism and the standpoint theory, they go beyond those epis-temologies (e.g. in terms of language, which according to the postmodern approach is not considered as a representation, a reflection of reality, but as its presentation, and therefore interpretation). Postmodern feminists (such as Jane Flax, Donna Haraway) are skeptical about the idea of com-mon awareness or the unity of women and their experiences; they reject universal (and universalizing) claims about the existence, character and power of reason. They also question the idea that there is such a thing as objective reality. According to Harding (1987b, pp. 186-187; 2008), post-modern theories are relevant to contemporary societies that are in a per-manent state of change and crisis.

In view of the previously assumed goal, in the later discussion I will primarily refer to the standpoint theory. It seems at this point the most adequate (most promising) approach for the reflection on the issues that I find interesting, and these are: experiencing social reality from the point of view of a woman and self-reflexivity of a female researcher.

Experience

Personal experience—in line with the thesis: *personal is political*—has gained the status of a significant source of knowledge in feminist research. Its value lies in the fact that from the most original level, the most primary one, it discloses what is important from the point of view of women. Rosi Braidotti (1995, p. 34) considers the experience, and actually the "lived experience of real-life women" as the central concept and the foundation of feminism. She discusses the category of 'experience' reaching to the term 'politics of location,' coined by Adrienne Rich (1984, pp. 210-231). "The politics of loca-tion means that the thinking, the theoretical process, is not abstract, univer-salized, objective, and detached, but rather that it is situated in the contin-gency of one's experience, and as such it is a necessarily partial exercise" (Braidotti, 1995, p. 34). However, this bias has a positive value because it promotes fuller presentation of the reality. Turning the attention to the ex-perience of women (specific, everyday, individual, personal, bodily), femi-nist researchers want to reach the key knowledge, namely the knowledge of women's position as resulting from a specific location in the social space. What I know about myself and others is *explicite* determined by my location within the society. Thus, from the perspective of feminism, social research is about the disclosure of the contexts of social activity, especially those areas that have been/are ignored, distorted or invisible in the discourse.

Braidotti (ibid., p. 35) draws the attention to another aspect linking the areas of 'experience' and 'location' of women. As she states, in terms of the feminist approach, the main habitat of location is the body: "The subject is not an abstract entity, but rather a material embodied one. The body is not a natural thing; on the contrary, it is a culturally coded socialized entity." The field of particular tension, where the experience and physicality of women are ousted, is provided by science. Dorothy Smith shows, on the example of sociology, that a woman who wants to deal with science is required to suspend the knowledge of herself and of her own sex, in other words, she needs to perform a separation of an intellectual, abstract 'I' from the fleshy, everyday 'I' (Smith, 1990, pp. 21-22). Such a situation causes a kind of split personality. To prevent this, we should not rub out our 'traces' permeating into the language, questions, problems and interpretations, but on the contrary—we ought to knowingly subject them to reflection, and incorporate as an important part of the research work. The category of 'experience' (personal and private) can therefore be regarded not only as a source of knowledge, but also as a method of cognition. As a researcher, I can build something of value in terms of cognitive science, not denying the experience of being a woman. Deliberate 'erasing' of what appears to be personal, emotional, confusing, unnecessary, unsightly, literary, etc., is *de facto* the act of 'blurring' the credibility of the description/picture of reality. Ironically, everything that goes beyond the canon of science in its positivist sense, is significant from the point of view of feminist methodology. When the description of the world is as close as possible to what is manifested in concrete experience, then—as said by Dorothy Smith (ibid., p. 62)—science is reliable. Its primary objective should be the reflection on its own position.

Reflexivity

Reflexivity is in the research procedure an essential tool to take care of the quality and transparency of the process. It is not irrelevant to the overall assessment of the objectivity and reliability of the research. For Donna Haraway (1988), objectivity (feministic one) means "situated knowledges." In her opinion, obtaining the 'full,' 'total,' 'all-encompassing' and the so-called neutral knowledge is unrealistic. "Knowledge from the point of view of the unmarked is truly fantastic, distorted, and irrational" (Haraway, 1991, p. 587). One cannot be at the same time in all positions, so only a partial/local perspective offers an objective look. Objectivity is always 'watching from somewhere'; it is not about transcendence and tearing apart the relationship between the subject and object in the study. It is knowledge resulting from looking from a particular social perspective; a look that does not claim to ensure absolute knowledge, but takes the responsibility for that part of reality which it sees and displays. This is the so-called strong objectivity, arising from the awareness of (one's own) location in the social space (Harding, 2004).

Jennifer Brayton argues that traditional social studies, and the postulate of objectivity preached by them, carry a 'flaw,' because they do not take into account the basic notion: they do not see how their own assumptions (prejudices, opinions, etc.) influence the research process, from the selection of the study subject, to the final presentation of the results (Brayton, 1997). An important consequence associated with a privileged position of a scientist-researcher is highlighted by Smith. In her view, a description/voice of a researcher is socially privileged; in the sense that it is spoken from the position of an authority, or someone to be listened to/to be heard, because his/her voice is loud (louder than the voices of others). Critique usually misses the fact that the voice of a scholar is not the only version of what is 'real,' 'true,' and 'objective,' but since the other voices are less audible (or even inaudible at all), they are not taken into account (Smith, 1990, p. 33)

According to the point of view of feminists, a research method that leads to acquiring reliable knowledge is the use of critical thinking. This reflection takes place in two ways: "It can mean reflecting upon, critically examining and exploring analytically the nature of the research process in an attempt to demonstrate the assumptions about gender [...] relations which are built into a specific project. It may also refer to understanding the 'intellectual autobiography' of researchers" (Maynard, 1995, p. 108). Reflecting upon the autobiographical dimension of the research work is significant because life experience of the researcher affects the analytical process and, eventually, the interpretation and the derived conclusions. In various feminist studies, according to Maynard, gender is not perceived just "as something to be studied, but as an integral dimension of the research process which itself is to be examined" (ibid.).

Language

The situation of conducting a study means for a female researcher directing special attention to the linguistic methods used for constructing and attributing sense to specific experience on the part of women. In order to make this possible, an essential element is not only the use of the language that is understood by both the informer and the researcher, but also the existence of such a language that allows women to express themselves in a manner acceptable to them. The problem is—according to Lucyna Kopciewicz (2003, p. 25)—the "non-existence of an alternate language that you can use to be more adequate (not referring to the existing meanings of femininity) while describing specific female experience."

Marjorie DeVault (1990, p. 96) notes that it is impossible to mirror the specificity of women's point of view without the simultaneous acquisition of the dominant language, which still remains the language of the male discourse. From a feminist point of view, the language used by men cannot be also the language of women, for instance because it ignores or devalues the experience of women; what is female is located outside the framework of rationality, logic or science. Within the discourse practiced by men, women's point of view, and thus also the way of speaking and writing characteristic for women, remains on the margins of what is valuable. Since they do not fit the categories typical for the male discourse, women's narratives appear to be chaotic (hysterical), incidental or even insignificant. This is the source of women's hardships in speaking and writing with their own voice; in expressing their own experience; and also with the free manifestation of "their own vision of the world, different from the male patriarchal one" (Ślęczka, 1999, p. 413). Hence, this is also the origin of the 'complex of a female author' and the absence or marginal position of women in traditional historiography. For feminism, it is the argument confirming the thesis that the traditional historical discourse makes a continuous story of the male experience. Women, to communicate their visions of the reality, to become 'audible,' must mimic male language, need tools for 'translation' or appropriate adaptive techniques, otherwise they are condemned to silence or being misunderstood. An interesting example of an adaptive technique of female authors is given by Virginia Woolf (1989 [1929], p. 74), who writes, "She met that criticism as her temperament dictated, with docility and diffidence, or with anger and emphasis", and further on: "She had altered her values in deference to the opinion of others" (ibid.). Woolf, on the example of literature and women's writing, shows that the values relevant from the point of view of women do not coincide with those valid for men, and that is why women—in order to reach positions of writers—were forced to adjust their visions to the requirements of the ruling authority.

Feminism offers a chance to 'recover' women's history. *Herstory* (or actually *Her story*)—the category created as a result of playing with the meanings grown from the word 'history' introduced by Adele Aldridge (Marzec, 2010, pp. 34-43)—is a way to run a narration from the point of view of women, a microperspective, allowing for analyzing female characters, through the story of what has been hitherto ignored (Kusiak, 1995). Natalie Zemon Davis implies that one can write history changing not only the object of study, but also the language of description (Świerkosz, 2010). What is likely to happen, therefore, is going beyond the definitions of woman in terms of negation or reversal of the male pattern. Feminist research pursues launching alternative, feminine discourses, and indirectly—liberating from the dominant, stereotypical patterns of speaking and writing about women and on their behalf. Monika Świerkosz writes, "What connects various feminist projects researching the past is the common goal—to resist the notion that women do not have their (significant) place in the history, which would make them beings devoid of historical consciousness, uprooted from traditions, disinherited from culture" (ibid.). A similar objective, referring to historical research, also guided the contemporary feminist research, both literary and social. The problem of their own language, the status of women and their presence in the official discourse still persists. The issue of the patriarchal horizon in women's thinking remains unresolved and it is passed on and perpetuated in the process of socialization and education. A challenge for feminist researchers, therefore, becomes giving women access to the space for free expression and (self)reflection, and for the recovery of *Herstory* There is a challenge and a need for writing about the experience of women, giving their accounts, their stories.

One more matter is undeniably important for female researchers. I mean here the awareness that the separation from the (male) language, as well as the methods of cognition and theories organized by the patriarchal rules, cannot be tantamount to abandoning science. It is therefore essential, being a feminist, not to exclude oneself, not to hide in hermetically sealed enclaves, available only to groups identifying themselves with the postulates of feminism.[17] The idea of modern science should not be reduced to a 'role reversal,' i.e. the substitution of the so-called male science with the female science. In feminism—as argued by Evelyn Fox Keller (1985, p. 3; quoted after Ślęczka, 1999, p. 432) (a representative of feminist empiricism)—it is all about "eliminating the absence of women in the history of social and political thought." To put it more generally, the feminist vision of science seeks to integrate different aspects of the human experience: both male and female. It involves not only tracking the mechanisms of discrimination/marginalization of women in science, but also active and systematical integration of their potential, knowledge and experience into science. The task of feminist researchers in this situation is broadening the field of science with views and reflections represented and established by women.

---

[17] It is true that within feminism there are voices that call for escapism from the world of men and, therefore, also from the science cultivated according to the androcentric perspective. For example, Mary Daly postulates the creation (by women) of a new living space designed exclusively for women.

Feminist Imagination

In the presented text, I attempted to meet two goals. First, I tried to articulate and organize the main themes of the feminist methodology, with the emphasis placed on the emanation of arguments that can be directly applied to the biographical studies of women's experiences. I figured *experience, reflexivity* and *language* as the most important categories. These categories I adopted as valid also for the level of analysis of the researcher's role and location in the research project. Second, my aim was to launch a process of self-reflection over the autobiographical dimension of my own research, which I link to—in this case—the (continuous) animation of 'feminist imagination.'

Adoption of the feminist perspective, which I support, is the choice opting for connecting the research and pedagogical-emancipation dimensions. The task of conducting feminist-oriented research involves not only documenting various aspects of the reality as experienced from the women's point of view, but it is also about taking a personal, political and committed approach to the world. A part of the pedagogical and emancipatory dimension of feminist works, requiring a strong emphasis, is the above-signaled feminist imagination (Bell, 1999).[18] This category I consider to be a special kind of self-awareness, to be precise—the awareness of one's own position in the social space, which is much more than simply the ability to analyze the social world (history, politics, etc.) in terms of 'difference' and 'gender.' Feminist imagination appears also as specific social sensitivity, manifested, *inter alia*, in the action for equality and fair treatment of people irrespective of their gender, age, and socio-ethnic origin. Feminist imagination is however, above all, courage; courage to make my life the central value; to subject it to creative self-reflection.

The topic proposed by me (autobiographical contexts of feminist studies) was just sketched here. It calls for further analysis and taking up in-depth studies, supported by further reading and critical reflection; however I already feel satisfied due to the fact that I dared to take the challenge of writing from the perspective of a woman. After years of searching, I discover in myself feminist imagination through which I gain new insights both into the research issues that I am interested in (biographical experiences of women) and my own life. In this context, the postulate of Gayatri Chakravory Spivak, calling for 'working on one's own ignorance' has a special ring to it (Spivak, 1990, p. 9). Further results of my work on my ignorance I hope to present in the near future.

---

[18]   It is worth recalling here the competences called 'sociological imagination' and 'anthropological imagination' known in human and social sciences.

References

Alexander, I. (1989). A Conversation on Studying and Writing about Women's Lives Using Nontraditional Methodologies. *Women's Studies Quarterly, 17* (3-4), 99-114.

Bell, V. (1999). *Feminist Imagination: Genealogies in Feminist Theory.* London – Thousand Oaks – New Delhi: Sage.

Braidotti, R. (1994). *Nomadic Subjects: Embodiment and Sexual Difference in Contemporary Feminist Theory.* New York: Columbia University Press.

Braidotti, R. (1995). Subject in Feminism. *Kwartalnik Pedagogiczny, 1-2* (155-156), 27-43.

Brannon, L. (2002). *Gender: Psychological Perspectives.* Boston: Allyn and Bacon.

Brayton, J. (1997). *What Makes Feminist Research Feminist? The Structure of Feminist Research within the Social Sciences,* www.unb.ca/par-l/win/feminmethod. htm [last accessed: May 25, 2012].

Chodorow, N.J. (1995). Gender as a Personal and Cultural Construction. *Signs, 20* (3), 516-544.

DeVault, M. (1990). Talking and Listening from Women's Standpoint: Feminist Strategies for Interviewing and Analysis. *Social Problems, 37* (1), 96-116.

Ellis, C. (2009). *Revision: Autoethnographic Reflections on Life and Work.* Walnut Creek: Left Coast Press.

Ellis, C. (2004). *The Ethnographic I: A Methodological Novel about Autoethnography.* Walnut Creek: AltaMira Press.

Ellis, C., & Bochner, A.P. (Eds.) (1996). *Composing Ethnography: Alternative Forms of Qualitative Writing.* Walnut Creek: AltaMira Press.

Graff, A. (2003). Feministki—córki feministek, czyli trzecia fala dobija do brzegu. In I. Kowalczyk & E. Zierkiewicz (Eds.), *W poszukiwaniu małej dziewczynki* (pp. 23-39). Wrocław: Stowarzyszenie Kobiet Konsola.

Haraway, D. (1988). Situated Knowledges: The Science Question in Feminism and the Privilege of Partial Perspective. *Feminist Studies, 14* (3), 575-599.

Harding, S. (2008). *Sciences from Below: Feminisms, Postcolonialities, and Modernities.* Durham – London: Duke University Press.

Harding, S. (2004). Rethinking Standpoint Theory: What is 'Strong Objectivity'? In. S. Harding (Ed.), *The Feminist Standpoint Theory Reader: Intellectual & Political Controversies* (pp. 127-140). London – New York: Routledge.

Harding, S. (1987a). Is There a Feminist Method? In S. Harding (Ed.), *Feminism and Methodology: Social Science Issues* (pp. 10-13). Bloomington – Indianapolis: Indiana University Press.

Harding, S. (1987b). Conclusion: Epistemological Questions. In S. Harding (Ed.), *Feminism and Methodology: Social Science Issues* (pp. 181-189). Bloomington and Indianapolis: Indiana University Press.

Harding, S. (1991). *Whose Science? Whose Knowledge? Thinking from Women's Lives.* Ithaka – New York: Cornel University Press.

Keller, E.F. (1985). *Reflections on Gender and Science.* New Haven: Yale University Press.

Kopciewicz, L. (2003). *Polityka kobiecości jako pedagogika różnic.* Kraków: Oficyna Wydawnicza Impuls.

Kusiak, A. (1995). O kobiecej historiografii. *Kwartalnik Pedagogiczny, 1-2* (155-156), 119-132.

Lipsitz Bem, S. (1993). *The Lenses of Gender: Transforming the Debate on Sexual Inequality.* New Haven: Yale University Press.

Marzec, L. (2010). Herstoria żywa, nie tylko jedna, nie zawsze prawdziwa. *Czas Kultury, 157* (5/2010), 34-43.

Maynard, M. (1995). Feminist Social Research. *Kwartalnik Pedagogiczny, 1-2* (155-156), 97-117.

Olesen, V. (2005). Early Millennial Feminist Qualitative Research: Challenges and Contours. In N.K. Denzin & Y.S. Lincoln (Eds.), *Handbook of Qualitative Research*, 3rd ed. (pp. 235-278). London – Thousand Oaks – New Delhi: Sage.

Ramazanoğlu, C., & Holland, J. (2002). *Feminist Methodology: Challenges and Choices*, London – Thousand Oaks – New Delhi: Sage.

Rich, A. (1986). *Of Woman Born: Motherhood as Experience and Institution.* New York – London: Norton.

Rich, A. (1984). Notes Towards a Politics of Location. In A. Rich, *Blood, Bread and Poetry: Selected Prose 1979-1985* (pp. 210-231). New York: Norton.

Richardson, L. (2005). *Writing: A Method of Inquiry.* In N.K. Denzin & Y.S. Lincoln (Eds.), *Handbook of Qualitative Research*, 2nd ed. (pp. 923-948). London – Thousand Oaks – New Delhi: Sage.

Smith, D.E. (1990). *The Conceptual Practices of Power: A Feminist Sociology of Knowledge.* Toronto: University of Toronto Press.

Spivak, G.C. (1990). Criticism, Feminism, and the Institution (Interview with Elizabeth Grosz). In S. Harasym (Ed.), *The Post-Colonial Critic: Interviews, Strategies, Dialogues* (p. 9). London – New York: Routledge.

Starr, C. (2000). "Third-Wave Feminism" [entry]. In L. Code (Ed.). *Encyclopedia of Feminist Theories* (p. 472). London – New York: Routledge.

Ślęczka, K. (1999). *Feminizm: Ideologie i koncepcje społeczne współczesnego feminizmu.* Katowice: Wydawnictwo Książnica.

Świerkosz, M. (2010). *Nie jestem siostrą mojej matki. Międzypokoleniowe dylematy feminizmu III fali,* http://www.unigender.org/?page=biezacy&issue=04&article=03 [last accessed: June 20, 2012].

Woolf, V. (1989 [1929]). *A Room of One's Own.* (M. Gordon, Pref.). San Diego – New York – London: A Harvest Book/Harcourt, Inc.

Chapter Nine

BIOGRAPHICAL SELF-MONITORING AS A CONDITION OF REFLEXIVE SOCIOLOGY AND ANTHROPOLOGY

TOWARDS PRACTICAL EPISTEMOLOGY OF SOCIAL SCIENCES

by Łukasz M. Dominiak

The editor of the present collection, while providing the future authors with a draft of a joint reflection on the 'professional' and 'non-professional' dimensions of humanistic experiences, raised an extremely important issue, he namely suggested a relationship between the practice of the profession and the biography of a humanist. We may well wonder whether the wording used by Marcin Kafar is adequate and if it brings us closer to new knowledge on the associations occurring between creative activity and its effects on the one hand, and the subjects causing them on the other. It is impossible, however, to negate the fact that at the time of the prevailing doubts concerning not only the sense of following the author and his/her work, but also the heuristic potential of the humanities, addressing the issue of coexistence of the 'author' and 'his/her work' is extremely important.

The task placed before us seems to be difficult, even due to the multiplicity and diversity of the emerging questions and concerns. Since, however, according to Pierre Bourdieu (1998), in no field of knowledge, which particularly refers to the social sciences, the progress of knowledge is attainable without taking into account the mechanisms governing the circumstances of cognition, the attempt aimed at showing the real position and function of the author in science certainly justifies the effort. And this is the goal, and also the starting point, of these deliberations. Immediately, I hasten to explain that my ambitions are rather contributory, than analytical.

Therefore, in the next sections, I will only briefly draw the attention of the scholars representing the disciplines included in the scope covered by the humanities and dealing with human communities (I mainly refer here to anthropology and sociology, which is also suggested by the title of this chapter), to a few autobiographical aspects of the knowledge generated by them. The basis of my argument will be the proposals outlined by various known and respected representatives of the contemporary humanities: Umberto Eco (born in 1932) and Michel Foucault (1926-1984).

For Eco (2002, p. 81) "The text is there" and "the private life of the empirical authors is in a certain respect more unfathomable than their texts" (ibid., p. 88). These terse assertions—uttered in a single breath by the author, who, as we know very well, has repeatedly manifested his uncommon skills that distinguish him in both the field of science and art—should be sufficient for all humanists as the validation of their reluctance towards (auto)biographical studies. However, the matter is not as simple as it looks at the first glance. The author of *Name of the Rose* eventually comes to the conclusion that the relationship between the text and the author does matter after all, and it is because of the subtle sensitization of the reader-recipient that follows it, which prepares him/her for making interpretations within a particular culture. The skillful use of the biographical signature in the text ensures the opportunity to counteract many unjustified interpretations, thanks to it the author is able to prevent the emergence of allusions that are going too far from his original intent. In such a case—even though we do not always know the exact set of meanings prepared by the author for his works (for whose reception he is rarely, if ever, able to take full responsibility)—we can define the basic text attributes, such as its recipients, the purpose of the statement, its style, etc. The creators of literature have few worries concerning it. As if by definition they do not need to bother about how the meanings package that they send to the anonymous recipient will be read—works seen from such angle resemble more the text that "is put in the bottle," mentioned by Eco (p. 67), which is sent to an unknown destination, than a 'brick' sited in the solid building of knowledge. Does the same refer to the texts contained in the area of the humanities? Who is the author in them and what are his role and responsibilities?

This question was answered by Michel Foucault in a short, but very frequently quoted text entitled *What is an Author?* (1969) and—indirectly—in the popular *The Order of Things* (2005 [1966]). In the second of these works, the French philosopher briefly presented the history of the human sciences, referring in this context to their contemporary state, conditioned by specific epistemological disposition. It is based on the fact that the humanities were not offered any finished concept of man from the outside, and they did not find themselves within the already defined epistemologi-cal field. This was not caused by any discovery, ideology, or any political option. According to Foucault, the existing state of affairs is the result of the crisis of representation, which caused a redistribution of the *episteme*, unac-companied by providing the access to the relevant body of knowledge, or even the ready set of discourses about man. The archaeology of knowledge reveals as well that, even assuming the perspective of the prehistory of the humanities, man himself, *a fortiori* authorizes any reflection on himself. In the opinion of Foucault, in the history of humanities there was not a single founding act, establishing the rules of a new set of discourses. A situation in which the humanities are a side effect of the inner workings of the whole area of knowledge does not bring about any reasons for isolation, or for the autonomy and autotelicity, but it radically and clearly shows the avail-able cognitive relationship. It is man—a given, historical, multi-dimensional character that is responsible for initiating the discourses and for the selec-tion of specific epistemological strategies (e.g. making an alliance between the humanities and the mathematical sciences or deriving from a particular philosophical tradition). In the analysis of the knowledge order advanced by the author of *The History of Sexuality*, man as a living being, working, talking and consuming became, at the turn of the eighteenth and nineteenth century, the subject of thought and knowledge, as well as the initial point for reflection establishing the thinking and knowledge about him—man as something that is equal to the thinking and knowledge of the celestial bod-ies, cells or languages (cf. Foucault, 2005 [1966], pp. 206-212).

What kind of man-author is Foucault talking about? Who is the subject of cognition in [A]nthropology for the philosopher, who largely contributed to the revolutionization of his traditional, substantial image? In the essay *What is an Author?* Foucault (1984 [1969], p. 111) presents the following, contained within the scope of literary criticism, definition: "the author provides the ba-sis for explaining not only the presence of certain events in a work, but also their transformations, distortions, and diverse modifications (through his biography, the determination of his individual perspective, the analysis of his social position, and the revelation of his basic design." This approach sat-urated with textuality, leads to highlighting the "author functions" (mean-ing the author as a function of discourse) entailing the reduction of the writer/creator/scientist to the role of the 'common denominator' of all his/her texts; the differences arising here should be abolished through the prin-ciples of evolution, influence, or maturation. Using such methical strat-egy, Foucault dismembers the author of any discursive statement into three parts, or, better to say, three modal "selves." They are not, though, fictional duplications: the first is the "I" speaking in the course of a demonstration; the second—the "self" providing a glimpse backstage of the effect of the work performed by the first "I"; the third is the "self" that tells about the achieved results and the problems emerging from the sum of the work performed by its predecessors. It should be emphasized that none of the separate "selves" refers in a simple way to the *real* author, and they only produce a kind of what we would call the 'speaking selves' (ibid., pp. 110-113).

The well-known concept in the theory of literature of the impersonal and biased subject, following the linguistic and autobiographical turns is also gradually absorbed by the social sciences. My statement is trying to fill a gap in the usage, on the part of the humanities, of the abundant benefits of the 'archaeological' discoveries of the French thinker.[1]

Staying for a moment within the diagnoses set by Foucault, I should also mention that modern *episteme* does not envisage any specific place for the social sciences taken in general, and for the humanities in particular. "The three faces of knowledge" do not include an object that could be dealt with by today's humanities, while maintaining by it at least relative autonomy from the philosophical reflection, as well as mathematical sciences and empirical sciences. The everyday life of humanities is turned inside out with a chronic, nebulous apportionment, moreover, it lacks a fixed, in-rooted location; it is an everyday life that remains in constant danger of extreme fragmentation. What does it mean? No less, no more, but the necessity to speedily put aside the old and in practice unresolved disputes (as to whether the humanities are/are not a science, and in what form (?); whether, and if so, to what extent do they require commitment (?); do they have the right to interfere in the society (?); do they require legitimacy (?); whether and to what extent are they literature (?); do they depend on politics (?), etc.), to reach an agreement as for the expectations with regard to the cognitive foundation, namely the paradox of anthropology (Sojak, 2004, pp. 110-116).

If the hitherto carefully concealed subjectivity of a subject were to become, even if only partially, something objectivized—I am still drawing inspiration from Foucault's conclusions—the components of identity of a subject and any activities related to them would have to be decoded in

---

[1] The dissonance between the saturation of the humanities with Foucault's works and actual taking into account the recommendations arising from them in the research and writing practice is all the more striking, the more strongly we realize not only how often the name 'Foucault' is invoked upon the occasion of various kinds of *excursus*, diagnoses and research, but also the importance of Foucauldian theorems for several weighty concepts currently valid in the social sciences, such as the critical theory, constructivism, Actor-Network Theory. Indeed, it is hard to find an area of the humanities or a humanist not admitting (consciously or unconsciously) to have intellectual relations to or inspirations from the author of *The Order of Things* (cf. Chapter Five in this book).

advance by splitting the 'author function.' This will make it possible to 'successfully' develop stories about the social world; as a result it will protect us 'from' instead of bringing us 'to' the absurdity. Reaching each time for the multiple selves that correspond to each particular discourse (whose modality will be marked in our awareness), we will make the consecutive steps on the way to, as it seems, *literaturization* that is necessary for the current human sciences. In this way, we can prove that this or that research decision is not only a strategy used in the game for new knowledge, but it actively establishes both the object and the subject of cognition. Locating the 'authors' of texts undoubtedly raises the chance to reach this dimension of the humanities, which is deposited in the 'folds' and 'crevices' of modern *episteme*. To be able to take full advantage of this, it is important—in addition to the meticulous fulfillment of Foucauldian recommendations as for clarifying which 'self' *tells*, which *reveals the inside story* (here we can recognize the undeniably useful function of introductions, prefaces, notes on plates, acknowledgments, etc.), and which 'self' *formulates comments and synthesizes the achievements* (afterword, possible continuations, diagnoses, recommendations)—to lead constant discernment in the changing positions across the field of social generation of knowledge. This active participation of humanists in constructing the metalevel of the first or higher degree for the disciplines they represent would prevent as well the sociologism that freely forces itself upon us ('everything depends on the context'), which, for anyone who wants to consciously practice humanities, is often the first and the last instance. It is a truism to repeat, but even Foucault, reluctant towards sociology, refreshes our memory in this regard saying that "the author function is linked to the juridical and institutional system that encompasses, determines, and articulates the universe of discourses" (Foucault, 1984 [1969], p. 113).

The operation of decoding the author function requires huge amount of analytical work—and it is inseparable from the need to identify the dominant tracks, picking out root metaphors, reflecting upon the language *usus*, forms of narratives and arguments, the nature of the most basic unit of things, namely statements (*énoncé* in the wording of Foucault), etc. However, only the undertaking of this scale gives a lively hope for the fulfillment of the project of Foucault's humanities 'without an author.' An author, in the process of writing, would create a space in which he would be gone, disappearing in a maze of thoughts and words (cf. Foucault, 1984 [1969]; 1972 [1969]). The process indicated here most probably will force the daredevils willing to give in to the spirit of change to face the necessity to rethink some, too tender, ideas concerning the humanities, but in return it will provide the humanities with the opportunity to leisurely drift in the trihedron of modern epistemology.

I will come back now to the starting point of my discussion quoting a sentence said by Beckett and cited by Foucault: "'What does it matter who is speaking,' someone said, 'what does it matter who is speaking'" (Foucault 1984 [1969], p. 101). The message in this, let's admit it, slightly fickle phrase—when applied to the current dilemmas of the humanities—reveals an apparent indifference to the sources of speaking/writing. This is not because the modern human sciences, following the example of literature and art, will soon lead to the 'death' of the author, but since this apparent indifference is—through the use of certain practices of academic writing and professed professional ethics—a rule of the social world of humanists. A question placed in such subsoil concerning the weightiness of a scientific ('professional') biography is a crucial question from the point of view of the future of humanistic sciences. Since we accept the bipolar tangle of the object and subject of cognition, it is impossible to deny one of the parties this 'epistemological equation' (through systematic 'forgetting,' 'who' is speaking) without the visible depletion of the effects of the research work. At this point, I can already attempt to explicitly state my main postulate: **Let's practice self-reflection, selecting for it the most natural means available—an escape from narcissism, practiced in parallel with refraining from the complete removal of ourselves from our own scientific work**.

It seems that very often humanists do not realize that the basic cognitive situation they have to face is *epistemocentrism* understood not only as a form of scholastic mistake, but also as an inevitable consequence of the specific position of a scholar. Pierre Bourdieu (1998), in order to face this situation so embarrassing for social sciences, suggested that a researcher should apply a three-stage "principle of objectivity." The first stage of the 'constant looking in the mirror' means taking into account the impact of our social position: the available values, attitudes, disposition; the second comprises an analysis of one's own location within the academic space, where the so-called participant objectivization occurs—henceforth constructing one's own statements at the level of uncommitted metareflection is out of the question; here also follows the attempt to control one's pre-assumptions, limitations and conditions; the third stage consists of the discovery of a significant characteristic of the researcher, i.e. his/her taking up social practices as an object of observation and analysis. Theory, and therefore also the theoretical approach, is combined with social distance which must be recognized and grasped.

The consequence resulting from the above is the need to provide for each report of the study—a report on the report, which brings together all the necessary information that is used by the recipient to establish the biographical and social location of the author—the only *real* author of the research report.

ය

The *real* author of the text, in order to reveal his 'authorial self,' should present at the end of the paper some information about his 'professional' biography. As a graduate of archeology and ethnology (studies at the turn of the twentieth and the twenty-first century) he experienced both the traditional and the postmodern variants of these disciplines with all the rewards and disadvantages resulting from it, that—becoming too impeding for him—were then all too easily invalidated in the era of a return to empiricism. The fact that the author got closer to sociology (about 8 years ago) embeds him in a more 'scientistic' and less 'literary' model of understanding human societies. An important empirical inspiration in this area has been provided by studies of the history of science, showing, *inter alia*, a complex discursive entanglement of scientists.

References

Bourdieu, P. (1998). *Practical Reason: On the Theory of Action*. (S. Farage et al., Trans.). Stanford: Stanford University Press.

Eco, U. (2002). Between Author and Text. In U. Eco et al. (S. Collini, Ed.), *Interpretation and Overinterpretation* (pp. 67-88). Cambridge: Cambridge University Press.

Foucault, M. (2005 [1966]). *The Order of Things: An Archeology of the Human Sciences*. London – New York: Routledge.

Foucault, M. (1984 [1969]). What is an Author? In P. Rabinow (Ed.), *The Foucault Reader* (pp. 101-120). New York: Pantheon Books.

Foucault, M. (1972 [1969]). *The Archeology of Knowledge and the Discourse on Language*. (A.M. Sheridan Smith, Trans.). New York: Pantheon Books.

Sojak, R. (2004). *Paradoks antropologiczny: Socjologia wiedzy jako perspektywa ogólnej teorii społeczeństwa*. Wrocław: Wydawnictwo Uniwersytetu Wrocławskiego.

NOTES ON CONTRIBUTORS

Marcin Maria Bogusławski

Assistant lecturer at the Department of Epistemology and Philosophy of Science, University of Łódź. He is interested in scientific pluralism, contemporary humanities, and social and political philosophy. Reviewer of "Journals Showcase."

Correspondence address:
Uniwersytet Łódzki, Katedra Epistemologii i Filozofii Nauki,
ul. Kopcińskiego 16/18, 90-232 Łódź, Poland

e-mail: martinmaria.bog@voila.fr

Łukasz M. Dominiak

Assistant professor at the Nicolaus Copernicus University in Toruń, ethnologist, archeologist and sociologist. Research interests: history, methodology and theory of social sciences. Member of the Polish Sociological Association and the Polish Scientific Association of Archaeologists.

Correspondence address:
Uniwersytet Mikołaja Kopernika w Toruniu, Instytut Socjologii,
ul. Fosa Staromiejska 1a, 87-100 Toruń, Poland

e-mail: lukasz@umk.pl

Magdalena Matysek-Imielińska

Assistant professor at the Institute of Cultural Studies, University of Wrocław, a sociologist, and cultural studies specialist, author of the following works (a selection): *Współczesne wspólnoty: Od polityki do estetyki* (*Contemporary Communities: From Politics to Aesthetics*) (2012), *W świecie Don Kichota: Między racjonalnością, rozsądkiem i rutyną: Rozważania wokół myśli Alfreda Schütza* (*In the World of Don Quixote: Between Rationality, Reason and Routine: Considerations about the Thought of Alfred Schütz*) (2012), *Kultura jako zorientowane na przyszłość wyzwanie rzucone teraźniejszości: O roli przyszłości w myśleniu o kulturze* (*Culture as a Future-oriented Challenge to the Present: On the Role of the Future in Thinking about Culture*) (2013). She is currently interested in the post-war Polish humanities oriented around the theoretical issues of culture, and looks there for concepts inspiring the creation of the cultural studies as a new academic discipline in the 1970s. She also started work on comparative studies of the Polish post-war thought about culture in the area of Marxism and British cultural studies.

Correspondence address:
Uniwersytet Wrocławski, Instytut Kulturoznawstwa,
ul. Szewska 50/51, 50-139 Wrocław, Poland

e-mail: magdalena.matysek-imielinska@uni.wroc.pl

Monika Modrzejewska-Świgulska

Assistant professor at the Chair of Pedagogy of Creativity at the Department of Educational Studies, University of Łódź. Research interests: pedagogy of creativity, educational factors of creativity development, levels of creativity; social context of creativity; sociocultural animation; biographical and narrative researches. Her publications mainly concern issues of egalitarian activity and ways of studying it.

Correspondence address:
Uniwersytet Łódzki, Katedra Badań Edukacyjnych,
ul. Pomorska 46/48, 91-408 Łódź, Poland

e-mail: momodrzejewska@gmail.com

Aneta Ostaszewska

Assistant professor at the Department of the Social Pedagogy at the Institute of Social Prevention and Resocialization, Warsaw University, sociologist and pedagogue. She is interested in the feminist methodology and autobiographical contexts of female identity. Within her interests an important place is occupied by the phenomenon of mythologization and popular culture myths. She wrote two books about the phenomenon of Michael Jackson—*Michael Jackson jako bohater mityczny: Perspektywa antropologiczna* (*Michael Jackson as a Mythical Hero: Anthropological Perspective*) and *Post Scriptum: Po śmierci Michaela Jacksona* (*Post Scriptum: After the Death of Michael Jackson*).

Correspondence address:
Uniwersytet Warszawski, Instytut Profilaktyki Społecznej i Resocjalizacji,
ul. Podchorążych 20, 00-721 Warszawa, Poland

e-mail: A.Ostaszewska@uw.edu.pl

Marta Songin-Mokrzan

Doctor of ethnology, postdoctoral researcher at the Department of Economic Sociology and Social Communication at the Faculty of Humanities, AGH—University of Science and Technology in Kraków. She carries out a research project entitled *Special Economic Zone as a Space of Realization of the Neoliberal Imaginary*, financed by the National Science Center (Poland). Research interests: theory and methodology of anthropology, committed anthropology, ethnography of the state, economic anthropology and feminist anthropology.

Correspondence address:
Akademia Górniczo-Hutnicza,
Katedra Socjologii Gospodarki i Komunikacji Społecznej,
ul. Gramatyka 8a, 30-071 Kraków, Poland

e-mail: martasongin@gmail.com

Andrzej Paweł Wejland

Associate professor, professor at the Chair of Contemporary Culture Theories and Studies at the Institute of Ethnology and Cultural Anthropology, University of Łódź, sociologist and cultural anthropologist, methodologist, author of the fol- lowing books: *Analiza logiczna interrogacji i jej zastosowania w badaniach społecznych* (*A Logical Analysis of Interrogation and Its Application to Social Research*) (1977), *Prestiż: Analiza struktur pojęciowych* (*Prestige: An Analysis of Conceptual Structures*) (1983), *Obrazy grup społecznych: Studium metodologiczne* (*Pictures of Social Groups: A Methodological Study*) (1991), co-editor of a handbook *Wywiad kwestionariuszowy: Analizy teoretyczne i badania empiryczne: Wybór tekstów* (*Questionnaire Interview: Theoretical Analyses and Empirical Research: Selection of Texts*) (1983) and author of the following works (a selection): *Transdyscyplinarne wątki w pracach metodolog- icznych Jana Lutyńskiego* (*Transdisciplinary Threads in Jan Lutyński's Methodological Works*) (1996), *Badawcze strategie zdrowego rozsądku, czyli skarby małego metodologa* (*Research Strategies of Common Sense, or Treasures of a Small Methodologist*) (1999), *Jak żegnać lokalne paradygmaty: O metodologii wywiadu i naukowych wspólnotach dyskursu* (*How to Say Farewell to Local Paradigms: The Methodology of Conducting Interviews and Scholarly Discoursive Communities*) (2004), *Wspólnota świadectwa: Charyzmaty- czne opowieści o uzdrowieniu* (*The Community of Testimony: Charismatic Narratives of Healing*) (2004), *Latem w parku: Epifanie codzienności* (*In a Park in the Summer: Epiphanies of the Everyday*) (2010), *Dyskurs i tożsamość: Opowieści we wspólnocie nau- kowej* (*Discourse and Identity: Stories in the Scientific Community*) (2010), *Miasto nocą: O poetyce nokturnu antropologicznego* (*A City at Night Time: On the Poetics of Anthro- pological Nocturne*) (2012), *Antropolog i pojęcie świadectwa: O niektórych pułapkach tu badaniu terenowym* (*Anthropologist and the Concept of Testimony: On Some Pitfalls in Fieldwork*) (2013). His main research interests in the recent years have included: transdisciplinary strategies in social sciences, metaphorical scenarios of science, anthropological narrativism and narrative anthropology, genres of anthropologi- cal narrative, community and narrative, anthropology of testimony, stories of ev- eryday urban life, anthropology of the everyday and its stories.

Correspondence address:
Uniwersytet Łódzki, Instytut Etnologii i Antropologii Kulturowej,
ul. Pomorska 149/153, 90-236 Łódź, Poland

e-mail: apwejland@uni.lodz.pl

REVIEWER
Wojciech J. Burszta

TRANSLATOR
Magdalena Machcińska-Szczepaniak

TYPESETTING AND COVER DESIGN
Agnieszka Okińska

Printed directly from camera-ready materials provided to Łódź University Press
by the Department of Educational Studies

PRINT AND SETTING
Quick Druk